STUDENT WORKBOOK
FOR JACOBS

MATHEMATICS
A Human Endeavor

Third Edition

STUDENT WORKBOOK
FOR JACOBS

MATHEMATICS
A Human Endeavor

Third Edition

SUSAN KNUEVEN WONG

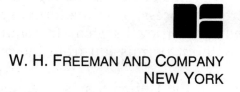

W. H. FREEMAN AND COMPANY
NEW YORK

ISBN 0-7167-2539-8

Printed in the United States of America

Second printing, 1999

Contents

4
Large Numbers and Logarithms 69

5
Symmetry and Regular Figures 91

6
Mathematical Curves 113

7
Methods of Counting 133

8

The Mathematics of Chance   149

9

An Introduction to Statistics   173

10

Topics in Topology   199

Appendices   219

Answers to Additional Exercises   281

Preface

This *Student Workbook* is designed as a complement to the textbook, *Mathematics: A Human Endeavor*, Third Edition, by Harold R. Jacobs. Although the textbook has enjoyed a remarkably successful history, this is the first edition for which W. H. Freeman and Company is publishing a workbook. It comes about as a result of the suggestions of many devoted *Mathematics* users who felt a workbook that provides spaces for answers to problem sets, graph paper, and other time-saving elements would enable students to better concentrate on the excellent presentation of the principles of mathematics in the textbook.

The workbook's author, Susan Knueven Wong, took the basic ideas of fellow devoted *Mathematics* users and added features that are either anticipated by, or inspired by, the textbook. The most important additional features are 86 short exercise sections. Students will find these exercises and the reiteration of many key ideas from the textbook beneficial in reinforcing and deepening their understanding of the textbook's topics.

The modular layout of the workbook allows instructors to pick and choose among its elements and features, which are described below. Features are divided between those drawn directly from the textbook and those designed as supplements to it.

FEATURES DERIVED FROM THE TEXTBOOK

♦ **Spaces for answers** are provided for all the textbook exercises. Sections are denoted by the headings SET I, SET II, and SET III. The workbook has been designed so that each lesson in the textbook is a separate and detachable unit of 2 or 4 pages, enabling instructors to assign individual lessons as in-class or homework assignments. Even if they are not used as assignments, they will make study and organization much easier.

♦ **Graph paper** is provided in Appendix A on pages 219–268. Within a lesson, both sides of a graph paper page should be used (except for exercises that call for graph paper to be cut up). Students should always begin new lessons on even-numbered graph pages so that the graph paper portions of answers can be separate and detachable. Enough extra graph paper has been provided to allow for some sides to be turned in unused, as well as to allow a student to start over on a fresh page if a mistake is made.

♦ **Textbook figures and tables** are reproduced in most instances where students are instructed to copy them to complete exercises. Most of the figures that the textbook directs students to trace are included. The figures that go along with the cut-out experiments in Chapters 5 and 10 are reproduced single-sided in Appendix B on pages 269–280 and may be cut out to suit the needs of the experiments. Other textbook figures are reproduced within lessons to help students visualize while working out their solutions. Students may find it useful to write on or around these figures, and thus *not* write directly into their textbook.

♦ A number of **key definitions and ideas** from the textbook are reproduced within lessons as reminders to help students. Their selection was decided by our workbook author, and to a large extent they reflect her experiences of areas where reminders might be helpful to students.

ADDITIONAL FEATURES

♦ **Additional exercises** are interspersed throughout the 67 lessons of the book's ten chapters. These 86 sections of short exercises were developed by Susan Knueven Wong over the course of many years of teaching from *Mathematics*. Instructors may assign them all, in part, or not at all. There are three different types of additional exercises:

- **Supplemental Exercises** are largely drawn from the subject matter presented in the lessons in which they appear. Successfully answering these exercises should enhance a student's understanding of the lessons as well as to build confidence in that understanding.

- **Reinforcing Past Lessons** are exercises based on the subject matter of previous lessons. They should remind students of prior lessons as well as to help keep fresh certain mathematical principles that will aid in the comprehension of subsequent lessons.

- **Using the Calculator** exercises apply the functions of graphing calculators to selected lessons from *Mathematics*. Graphing calculators can provide math students with challenging possibilities and have proven popular in courses at this level. They enable students to enter functions and visualize curves. Used effectively, they can further the student's visual perception of mathematics and reinforce the textbook topics. The exercises offered here directly connect the textbook to these useful instruments. They are not intended as a full curriculum, but rather as a starting point for instructors who want to integrate graphing calculators into this course. Any graphing calcultor can be used for these exercises. The author uses and recommends the TI-81 and TI-82 made by Texas Instruments.

♦ **Answers** to Additional Exercises appear in the back of the workbook on pages 281–000. Not all of the answers to the Using the Calculator exercises are included. Some of the screens are not reproduced.

♦ **Hints** to better understanding lessons and to solving exercises are offered at various points throughout the workbook. Like the key definitions and ideas reprinted from the textbook, they reflect areas where our workbook author's experiences suggest a bit of help could benefit some students. Most often hints appear at the beginning of a lesson, or in italics and parentheses within exercises.

The author and W. H. Freeman and Company welcome the comments and suggestions of students and instructors who use this workbook. Please send any remarks to:

Supplements Editor
W. H. Freeman and Company
41 Madison Avenue
New York, NY 10010

From the Author

I would like to thank those who have been helpful to me in the preparation of this workbook. I include my editor, Patrick Shriner, as well as Michele Barry and Erica Seifert at W. H. Freeman and Company. I would also like to thank Harold R. Jacobs for writing his wonderfully lucid textbook *Mathematics: A Human Endeavor*—a book that I have used with great success in my classes for many years. I felt it was an honor to be asked by W. H. Freeman to offer my interpretation of a workbook for this textbook. I hope that my efforts will meet with some degree of success and satisfaction among the students and instructors using it, and also that it will serve to increase the number of people exposed to this outstanding textbook.

S. K. W.
Indianapolis, Indiana
July 1994

Name _____

Date _____

Supplemental Exercises

A *ratio* is a comparison of two numbers by division.

The "ratio" of the number x to y is the number $\frac{x}{y}$.

Rectangles are *similar* when the ratios of the lengths to the widths are the same.

For example:

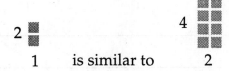

2 ▮   4 ▦

1   is similar to   2

because the ratio of $\frac{2}{1}$ is the same as the ratio of $\frac{4}{2}$.

Reduce these ratios to lowest terms:

1. $\frac{27}{45}$ = —   2. $\frac{48}{60}$ = —   3. $\frac{65}{91}$ = —   4. $\frac{96}{100}$ = —

SET I EXERCISES

1. Use the graph paper provided.

2. _____   3. _____   4. _____

5. _____

_____

6. Use the graph paper provided. Write the dimensions you chose along the side of each table.

7. _____

_____

8. Use the graph paper provided.

9. _____

_____

10. Use the graph paper provided.

11. _____

   _____

SET II EXERCISES

1. _____

2. Graph on the paper provided.   3. _____

4. $\frac{14}{4}$ = —   5. Graph on the paper provided.

6. _____ ; _____

   _____

7. $\frac{12}{10}$ = —   8. Graph on the paper provided.

9. _____ ; _____

   _____

10. _____

11. _____ ; _____

12. _____ ; _____

13. _____ ; _____

SET III EXERCISES

1. Graph on the paper provided.

2. _That you can draw lines in boxes_

3. _____

4. _____ ; _____

5. _____ ; _____

6. _____ ; _____

SET I EXERCISES

1. _____

2. _____ 3. _____ 4. _____

5. $1 \times 9 + 2 =$ _____

6. $12 \times 9 + 3 =$ _____

7. $123 \times 9 + 4 =$ _____

8. $1{,}234 \times 9 + 5 =$ _____

9. _____ $\times$ __ $+$ __ $=$ _____ 10. _____

11. _____ $\times$ __ $+$ __ $=$ _____ 12. _____

13. _____ 14. _____

15. _____ 16. _____

17. _____

18. _____

SET II EXERCISES

1. Number of points connected   2   3   4   5   6
   Number of regions formed   2   4   8 __ __

2.   Choose 5 points   Choose 6 points

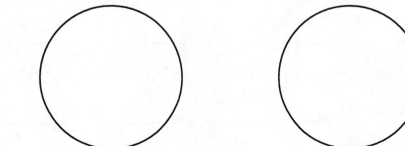

3. _____ ; _____ 4. _____

Use the graph paper provided to complete Exercises 5–17.

5. _____

6. _____ 7. Use graph paper.

8. _____ 9. Use graph paper.

10. _____ 11. _____ 12. Use graph paper.

13. _____ 14. Use graph paper. 15. _____

16. _____

17. _____

SET III EXERCISES

1. _____

2. _____ 3. _____ 4. _____

Reinforcing Past Lessons

1. Complete the table below for a general rule in terms of odd and even billiard table dimensions.

Corner ball ends up in	Length	Width
Upper right	Odd	Odd
Upper left	_____	_____
Lower right	_____	_____
Lower left	_____	_____

2. Summarize the table above by answering the following questions:

a. If the dimension of one side is even, then the ball ends up in which possible corners?

_____

b. If neither dimension is even, then the ball ends up in which possible corners?

_____

SET I EXERCISES

Deductive reasoning is a method of using logic to draw conclusions from statements we accept as true. Deductive reasoning is reasoning from the general to the particular.

A *theorem* is a statement that is proved by deductive reasoning.

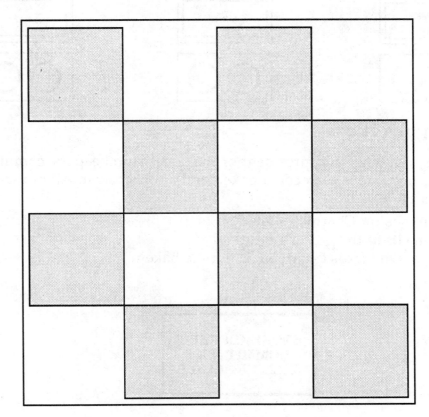

1. _____

2. _____   3. _____

4. _____

5. _____

6. Brown: _____ White: _____   7. Brown: _____ White: _____

8. Brown: _____ White: _____   9. _____

10. _____ ; _____

11. _____ ; _____   12. _____

13. _____

14. _____   15. _____

16. _____

17.   18.   19.

| 1 red 1 white | 2 white | 2 red |

SET II EXERCISES

Andre is a butcher and president of the street storekeepers' committee, which also includes a grocer, a baker, and a tobacconist. All of them sit around a table.

Andre sits on Charmeil's left.
Berton sits at the grocer's right.
Duclos, who faces Charmeil, is not the baker.

1.

STOREKEEPERS'
COMMITTEE
MEETING TABLE

(Use the table below to help you determine each person's occupation. Make a checkmark when you know an occupation is true, an "x" mark when you know an occupation is false. The marks for the statement "Andre is a butcher" have been provided.)

	Butcher	Grocer	Baker	Tobacconist
Andre	✔	✗	✗	✗
Berton	✗			
Charmeil	✗			
Duclos	✗			

2. Andre: _____   Berton: _____
   Charmeil: _____   Duclos: _____

3. _____   4. _____   5. _____

(In Exercises 6–21 remember that three faces of each large cube are hidden from view.)

6. _____ ; _____

7. _____   8. _____ ; _____

9. _____   10. _____ ; _____

11. _____   12. _____ ; _____

13. _____   14. _____

15. _____

16. ___ ; _____   17. ___ ; _____

18. ___ ; _____   19. ___ ; _____

20. ___ ; _____

21. _____

22. _____ ; _____

23. _____ ; _____

24. _____ ; _____

SET III EXERCISES

1. _____   2. _____

3. _____

Supplemental Exercises

Euclid, the ancient Greek mathematician mentioned in the introduction to this lesson, summarized many fundamental discoveries made by others.

1. One of Euclid's important theorems was "odd plus odd equals even." Write two numerical examples of this theorem.

   _____   _____

2. Write two numerical examples of his theorem "even plus odd equals odd."

   _____   _____

Using the Calculator

Pythagorean Theorem: $a^2 + b^2 = c^2$ where a and b are the legs of a right triangle and c is the hypotenuse.

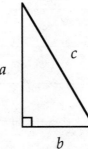

Use your calculator to determine the missing parts of the right triangles in Exercises 1–4. Show your work and equations. Draw each triangle with a straight edge. If the answers are not whole numbers, round to the nearest tenth.

1. Leg = 5, Hypotenuse =13

   Leg = _____

2. Legs = 4 and 7

   Hypotenuse = _____

3. Legs = 4.5 and 6.3

   Hypotenuse = _____

4. Legs = 16.5 and 13.2

   Hypotenuse = _____

Name _____

Date _____

Supplemental Exercises

The distributive rule is illustrated by the figures below.

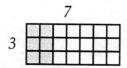

The number of the small squares in the figure can be expressed as either

$$3 \times (2 + 5) \text{ or } (3 \times 2) + (3 \times 5).$$

Notice that

$$3 \times (2 + 5) = 3 \times 7 = 21$$

and

$$(3 \times 2) + (3 \times 5) = 6 + 15 = 21.$$

Use the distributive rule to restate the following expressions.

1. $5 \times (2 + 4) =$ _____

2. $8 \times (4a + 11b) =$ _____

3. $0.5 \times (4a + 12) =$ _____

4. $12 \times (1 + 6b) =$ _____

SET I EXERCISES

1. _____   2. _____

3. _____   4. _____

5. _____   6. _____

7. _____   8. _____

9. _____   10. _____

Use the same format as the example below to complete Exercises 11–13.

	Examples		Proof 1	Proof 2
Choose a number.	5	8	□	n
Multiply by 3.	15	24	□ □ □	$3n$
Add 6.	21	30	□ □ □ oooooo	$3n + 6$
Divide by 3.	7	10	□ oo	$n + 2$
Subtract the number first thought of.	2	2	oo	2
The result is 2.				

	Examples	Proof 1	Proof 2
11. Choose a number.			
Add 3.			
Multiply by 2.			
Add 4.			
Divide by 2.			
Subtract the number first thought of.			
The result is 5.			

	Examples	Proof 1	Proof 2
12. Choose a number.			
Double it.			
Add 9.			
Add the number first thought of.			
Divide by 3.			
Add 4.			
Subtract the number first thought of.			
The result is 7.			

	Examples	Proof 1	Proof 2

13. Choose a number.

Triple it.

Add the number one larger than the number first thought of.

Add 11.

Divide by 4.

Subtract 3.

The result is the original number.

SET II EXERCISES

Trick 1	Example	1.
Choose any three-digit number.	524	_____
Multiply it by 7.	$524 \times 7 = 3{,}668$	$\_\_\_ \times 7 = \_\_\_\_$
Multiply the result by 11.	$3{,}668 \times 11 = 40{,}348$	$\_\_\_\_ \times 11 = \_\_\_\_\_$
Multiply the result by 13.	$40{,}348 \times 13 = 524{,}524$	$\_\_\_\_\_ \times 13 = \_\_\_\_\_$

2. _____

3. _____   4. _____

5. _____ $\times\ 1{,}001 =$ _____   6. _____

7. _____

Trick 2	8.	9.
Choose any two-digit number.	_____	_____
Multiply it by 13.	$\_\_\_ \times 13 = \_\_\_\_$	$\_\_\_ \times 13 = \_\_\_\_$
Multiply the result by 21.	$\_\_\_\_ \times 21 = \_\_\_\_\_$	$\_\_\_\_ \times 21 = \_\_\_\_\_$
Multiply the result by 37.	$\_\_\_\_\_ \times 37 = \_\_\_\_\_$	$\_\_\_\_\_ \times 37 = \_\_\_\_\_$

10. _____

11. $13 \times 21 \times 37 =$ _____

12. _____

SET III EXERCISES

Old "Magic" Trick	1.	2.
Write down the year of your birth.		
Write down the year in which an important event in your life occurred.		
Write down your age as of December 31, this year.		
Write down the number of years since the important event occurred.		
Add the four numbers together.		

3. _____

4. _____

   _____

Supplemental Exercises

The following is another logic exercise similar to the storekeepers exercise in Lesson 5, Set II. As with that exercise, you are provided with a list of statements you can accept as true. Use the table beneath the statements to help you keep track of the conclusions you draw using deductive reasoning.

Anthony, Ricardo, Marie, and David are all musicians. The instruments they play (though not in this order) are the saxophone, the piano, the drums, and the guitar.

The saxophone player is a male.
Neither Ricardo nor David plays an instrument with strings.
Marie and David *always* play seated.

	Saxophone	Piano	Drums	Guitar
Anthony				
Ricardo				
Marie				
David				

Who plays what instrument?

Anthony: _____   Ricardo: _____

Marie: _____   David: _____

SET I EXERCISES

1. $16 = \underline{\quad} + \underline{\quad}$

   $18 = \underline{\quad} + \underline{\quad}$

   $20 = \underline{\quad} + \underline{\quad}$

   $22 = \underline{\quad} + \underline{\quad}$

   $24 = \underline{\quad} + \underline{\quad}$

   $26 = \underline{\quad} + \underline{\quad}$

   $28 = \underline{\quad} + \underline{\quad}$

   $30 = \underline{\quad} + \underline{\quad}$

   $32 = \underline{\quad} + \underline{\quad}$

   $34 = \underline{\quad} + \underline{\quad}$

● 2. _____

3. _____

4.

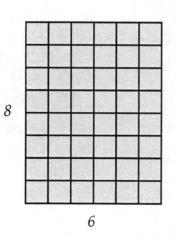

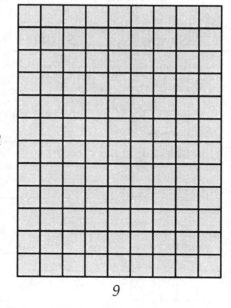

● 5. _____

6. _____

7. _____   8. _____

9. _____   10. _____

11. _____

12. _____

13. _____

14. _____

SET II EXERCISES

1. _____   2. _____   3. _____

4. _____   5. _____

6.

Piece	Shape	Squares it would cover
A		2 brown and 2 white
B		_____
C		_____
D		_____
E		_____

7. _____

Another Number Trick	Example	8.
Choose any three-digit number whose first and last digits differ by more than one.	175	
Write down the same three digits in reverse order to form another three-digit number.	571	
Subtract the smaller of your two numbers from the larger and circle your answer.	571 − 175 ⟨396⟩	
Reverse the digits in the number you circled to form another number and circle it.	⟨693⟩	
Add the two circled numbers together.	396 + 693 1,089	

9. _____

10. _____

Variation of Number Trick	Example	11.
Choose any three-digit number whose first and last digits differ by *exactly* one.	483	
Write down the same three digits in reverse order to form another three-digit number.	384	
Subtract the smaller of your two numbers from the larger and circle your answer.	483 − 384 ⟨99⟩	
Reverse the digits in the number you circled to form another number and circle it.	⟨99⟩	
Add the two circled numbers together.	99 + 99 198	

12. _____

13. _____

(As with the previous logic puzzles, you may find the table below useful in keeping track of your conclusions.)

Miss Green, Miss Black, and Miss Blue are out for a stroll together. One is wearing a green dress, one a black dress, and one a blue dress.

"Isn't it odd," says Miss Blue, "that our dresses match our last names, but not one of us is wearing a dress that matches her own name?"

"So what?" said the lady in black.

	Green dress	Black dress	Blue dress
Miss Green			
Miss Black			
Miss Blue			

14. What color is each lady's dress?

Miss Green: _____

Miss Black: _____

Miss Blue: _____

SET III EXERCISES

1. _____

2. _____

3. Draw a solution below.

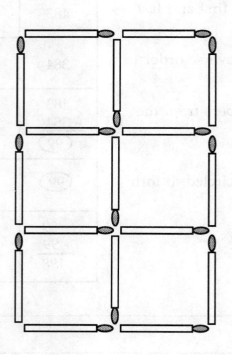

4. _____

Name _____

Date _____

Supplemental Exercises

To find the *common ratio* of a geometric sequence, divide the second number by the first number in a sequence (or the latter by the former).

For example, for the sequence

    2 6 18 54

the common ratio would be

$$\frac{6}{2} = 3, \text{ or } \frac{18}{6} = 3, \text{ or } \frac{54}{18} = 3.$$

This is expressed in another format as

$$\begin{array}{cccc} 2 & 6 & 18 & 54 \\ \nearrow & \nearrow & \nearrow & \\ \times 3 & \times 3 & \times 3 & \end{array}$$

Find the common ratio for each sequence.

1.	3	21	147	1029	____
2.	1	6	36	216	____
3.	192	48	12	3	____

SET I EXERCISES

1. _____

2. _____

3. _____

4. _____ _____ _____   5. _____   6. _____ _____ _____

7.   4   12   36   _____ _____

8.   11   121   _____ _____

9.   5   20   _____   320   _____

10. _____ _____ 72 144 288

11. 0 _____ _____ _____

12. 16 40 100 _____ _____

13. 81 54 _____ _____ _____

(Remember that in an arithmetic sequence you add the same number (common difference) to each term to get the next term. In a geometric sequence you multiply each term by the same number (common ratio) to get the next term.)

14. _____ ; _____ 15. _____ ; _____

16. _____ ; _____ 17. _____ ; _____

18. _____ ; _____ 19. _____ ; _____

20. _____ ; _____ 21. _____ ; _____

22. _____

23. _____ _____ _____ _____ 440 880 _____ _____

24. _____

25. _____

26. 960 720 _____ _____

SET II EXERCISES

1. _____ 2. _____

3. _____ 4. _____

5. _____ 6. _____

7. _____

8. t_4 $2 + 3 + 3 + 3$ $= \_\_\_ + \_\_\_ \cdot \_\_\_ = 11$

    t_5 $2 + 3 + 3 + 3 + 3$ $= \_\_\_ + \_\_\_ \cdot \_\_\_ = 14$

9. t_4 $2 \cdot 3 \cdot 3 \cdot 3$ $= \_\_\_ \cdot \_\_\_ = 54$

    t_5 $2 \cdot 3 \cdot 3 \cdot 3 \cdot 3$ $= \_\_\_ \cdot \_\_\_ = 162$

10. t_{10} = ___ + ___ · ___

11. _____

12. t_{10} = ___ · ___

13. _____

14. _____ _____ _____ _____ _____

15. t_{100} = _____

16. t_{11} = ___ · ___

17. t_{50} = ___ · ___

18. t_{123} = ___ · ___

19. t_8 = ___ · ___

20. t_n = ___ · ___

Supplemental Exercises

1. Use the information you learned in Set II to evaluate the following geometric sequences. Show all the steps to your work.

    a. Find the 5th term when the 1st term is 3 and the common ratio is 2.

        _____ = _____

    b. Find the 6th term when the 1st term is 4 and the common ratio is $\frac{1}{2}$.

        _____ = _____

    c. Find the 4th term when the 1st term is 6 and the common ratio is 3.

        _____ = _____

2. The growth of the population of the world can be gauged as a geometric sequence for the (approximate) years that are noted below.

    a. Complete the sequence.

Year	400	1650	1850	1930	1975
Pop.	250,000,000	_____	1,000,000,000	_____	_____

    b. What is the common ratio of the sequence? _____

SET III EXERCISES

1. _____   2. _____   3. _____

4. _____   5. _____   6. _____

7. _____

8. _____

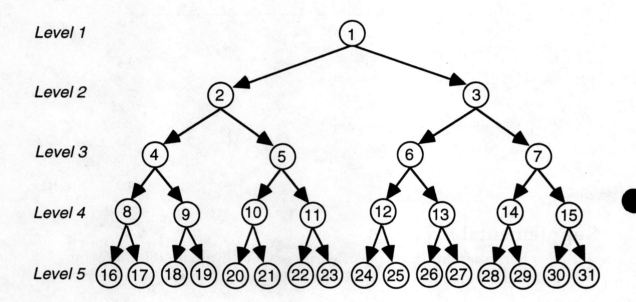

CHAPTER **2**

LESSON 3

Name _____

Date _____

SET I EXERCISES

A *binary sequence* is a geometric sequence in which the first term is 1 and the common ratio is 2.

1. _____   2. _____

3. _____ _____ _____ _____ _____

4. 5 = _____   5. 20 = _____

6. 42 = _____   7. 71 = _____

8. 95 = _____

Number	Binary numeral						
	64	32	16	8	4	2	1
19*			1	0	0	1	1
100*	1	1	0	0	1	0	0

* 19 and 100 are offered as examples.

9. 5 ___ ___ ___ ___ ___ ___ ___

10. 20 ___ ___ ___ ___ ___ ___ ___

11. 42 ___ ___ ___ ___ ___ ___ ___

12. 71 ___ ___ ___ ___ ___ ___ ___

13. 95 ___ ___ ___ ___ ___ ___ ___

14. _____ 15. _____ 16. _____

17. ___ ___ ___ ___ ___
   1   0   1   0   0   _____

18. ___ ___ ___ ___ ___ ___
   1   0   1   0   0   0   _____

19. ___ ___ ___ ___ ___ ___ ___
   1   1   0   1   0   0   1   _____

20. ___ ___ ___ ___ ___ ___ ___ ___
   1   1   0   1   0   0   1   0   _____

29

21. _____

22. _____   23. _____

Supplemental Exercises

Convert these decimal numbers into binary numerals.

Decimal value	64	32	16	8	4	2	1
1. 56	___	___	___	___	___	___	___
2. 99	___	___	___	___	___	___	___
3. 68	___	___	___	___	___	___	___
4. 12	___	___	___	___	___	___	___

Convert these binary numerals into decimal numbers.

5. 1   1   0   1   1   _____

6. 1   1   1   1   1   1   0   _____

7. 1   0   0   0   1   1   _____

8. 1   0   0   1   _____

SET II EXERCISES

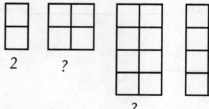

2   ?

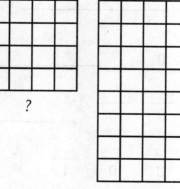

?   ?

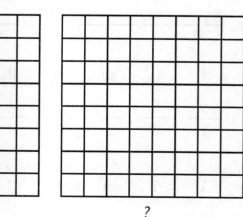

?   ?

1

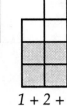
1 + 2 = ?

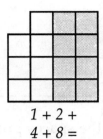
1 + 2 + 4 = ?

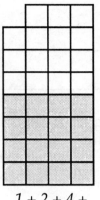
1 + 2 + 4 + 8 = ?

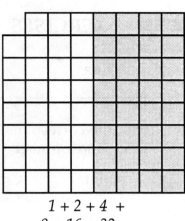

1 + 2 + 4 + 8 + 16 = ?

1 + 2 + 4 + 8 + 16 + 32 = ?

1.   2   ___ ___ ___ ___ ___

2.   1   ___ ___ ___ ___ ___

3.   _____

4.

t_1	t_2	t_3	t_4	t_5	t_6	t_7	t_8	t_9	t_{10}	t_{11}	t_{12}
1	2	4	8	16	32	64	___	___	___	___	___

5.   _____

6.   _____

7.   _____   8.   _____

9.   _____

10.   _____   11.   _____   12.   _____

13.   _____   14.   _____   15.   _____

16.   _____   17.   _____

18.   _____   19.   _____

SET III EXERCISES

1. _____

2. _____ ; _____

3. Draw the corresponding hexagrams in the spaces provided between their names and number values.

K'uei	Chia jên	I	Sung	Lü
10	20	30	40	50

Supplemental Exercises

Now that you know the formula for finding a sum of binary numerals, find the sums of the following. Do so without adding.

1. The sum of the first four terms.   _____

2. The sum of the first eight terms.   _____

3. The sum of the first eleven terms.   _____

Name _____

Date _____

SET I EXERCISES

The *counting numbers* are the numbers 1, 2, 3, 4, 5, 6, ... The *sequence of squares* is the set of the counting numbers taken to the second power. A *square root* of a given number, when multiplied by itself, gives the original number.

For example: The square root of 16 (written as $\sqrt{16}$) = 4, because $4 \times 4 = 16$.

1.   1   4   9   16   25   36   __ __ __ __ __ __
     ∨ ∨ ∨ ∨ ∨ ∨ ∨ ∨ ∨ ∨ ∨

2.   3   __ __ __ __ __ __ __ __ __ __

3.   _____

4.   _____

5.   _____ _____ _____ _____ _____
(This sequence is the total for each second, counting from the beginning.)

6.   _____

7.   _____ _____ _____ _____ _____
(This sequence is the amount traveled in each individual second.)

8.   _____   9.   _____

10.   4   8   _____ _____ _____ _____

11.   _____

12.   1   4   _____ _____ _____ _____

13.   _____

14.   _____ _____ _____ _____

15.   _____

16.   _____ _____ _____ _____

17.   _____

18. _____ _____ _____ _____ _____ _____ _____ _____ ___

19.
$$1 = 1$$
$$1 + 2 = \underline{\hspace{1cm}}$$
$$1 + 2 + 3 = \underline{\hspace{1cm}}$$
$$1 + 2 + 3 + 4 = \underline{\hspace{1cm}}$$
$$1 + 2 + 3 + 4 + 5 = \underline{\hspace{1cm}}$$

20. _____

21.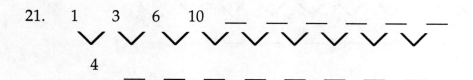

22. _____

23. _____
    _____

SET II EXERCISES

No.	Square	No.	Square	No.	Square	No.	Square
1	1	11	121	21	441	31	961
2	4	12	144	22	484	32	1,024
3	9	13	169	23	529	33	1,089
4	16	14	196	24	576	34	1,156
5	25	15	225	25	625	35	1,225
6	36	16	256	26	676	36	1,296
7	49	17	289	27	729	37	1,369
8	64	18	324	28	784	38	1,444
9	81	19	361	29	841	39	1,521
10	100	20	400	30	900	40	1,600

1. _____ 2. _____ 3. _____

4. _____ 5. _____ ; _____

6. _____ ; _____

7. _____ ; _____

8. _____ ; _____

9.

No.	Square	No.	Square	No.	Square
51	_____	54	_____	57	_____
52	_____	55	_____	58	_____
53	_____	56	_____	59	_____

10. _____ ; _____

11.   1   4   9   6   5   ___ ___ ___ ___

12. _____

13.

Number	Square	Digital root of square	Number	Square	Digital root of square
1	1	1	6	___	___
2	4	4	7	___	___
3	9	9	8	___	___
4	16	7	9	___	___
5	___	___	10	___	___

(Remember that the digital root of a number is always a single *digit.)*

14.

Number	Square	Digital root of square	Number	Square	Digital root of square
11	___	___	16	___	___
12	___	___	17	___	___
13	___	___	18	___	___
14	___	___	19	___	___
15	___	___	20	___	___

15. _____ ; _____

16. _____

SET III EXERCISES

1.

Subgroups	s	p	d	f
Number of elements	2	___	___	___

2. _____

3.

Row	1	2	3	4	5	6
Number of elements	2	8	___	___	___	___

4.
Row	1	2	3	4	5	6
Half of number of elements	1	4	___	___	___	___

5. _____

6. _____ 7. _____

Reinforcing Past Lessons

1. A woman who smokes an average of 34 cigarettes a day decides she wants to quit. Her doctor advises her to cut back on her daily average by four cigarettes each week until she gets down to zero.

a. Write out the sequence of her average number of cigarettes per day each week, assuming that she achieves the goals suggested by her doctor.

_____

b. What kind of sequence is this? _____

2a. Complete the next three terms of the geometric sequence below.

3   6   ___   ___   ___

b. What is the common ratio of this sequence? _____

Using the Calculator

Find the square root of each of the following terms using your calculator.

1. 3364 _____   2. 729 _____

3. 10,201 _____   4. 179,776 _____

Find the squares of each of the following terms using your calculator.

5. 191 _____   6. 3,030 _____

Name _____

Date _____

SET I EXERCISES

The *volume* of a rectangular solid is found by multiplying the length times the width times the height (l × w × h). Units for volume are "cubic."

In a cube, all three of these dimensions are the same, so the formula is side times side times side (s × s × s, or s³).

1.   1   8   27   64   125   ____ ____ ____ ____ ____

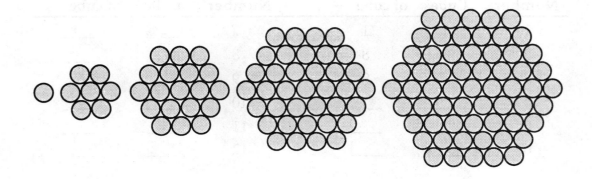

2.   1   7   ____ ____ ____

3. _____ 4. _____ 5. _____

6. _____

7.   6   24   ____ ____ ____

8.   1   8   ____ ____ ____

9. Length of edge   1   2   4   8   16
    Surface area   6   24   __   __   __
    Volume   1   8   __   __   __

10. _____

11. _____

12. _____

13. 1^4 $\quad$ 2^4 $\quad$ 3^4 $\quad$ 4^4 $\quad$ 5^4 $\quad$ 6^4

   1 $\quad$ ____ $\quad$ ____ $\quad$ ____ $\quad$ ____ $\quad$ ____

14. 1^4 $\quad$ 2^4 $\quad$ 3^4 $\quad$ 4^4 $\quad$ 5^4 $\quad$ 6^4

   ____ $\quad$ ____ $\quad$ ____ $\quad$ 16^2 $\quad$ ____ $\quad$ ____

15. _____

16. 1^7 $\quad$ 2^7 $\quad$ 3^7 $\quad$ 4^7 $\quad$ 5^7

   1 $\quad$ ____ $\quad$ ____ $\quad$ ____ $\quad$ ____

17.

Number	Cube	Digital root of cube	Number	Cube	Digital root of cube
1	1	1	7	____	____
2	8	8	8	____	____
3	27	9	9	____	____
4	____	____	10	____	____
5	____	____	11	____	____
6	____	____	12	____	____

18. _____

Supplemental Exercises

Find the digital root for the following numbers.

1. 3,458,978 _____
2. 88,888,000 _____
3. 45,457 _____
4. 123,456,789 _____
5. 7^4 _____
6. 15^2 _____

SET II EXERCISES

Remember that the sequence of cubes is the set of counting numbers taken to the third power.

1. 1 $\qquad$ = 1

   3 + 5 $\qquad$ = ____

   7 + 9 + 11 $\qquad$ = ____

   13 + 15 + 17 + 19 $\qquad$ = ____

2. _____ $\qquad$ 3. _____

4. ___ + ___ + ___ + ___ + ___ = ___

    ___ + ___ + ___ + ___ + ___ + ___ = ___

5. _____

6.   1   =   1

    1 + 8   = ___

    1 + 8 + 27   = ___

    1 + 8 + 27 + 64   = ___

7. _____   8. _____

9. 1^3 = 1^2

    ___ + ___   = ___

    ___ + ___ + ___   = ___

    ___ + ___ + ___ + ___   = ___

10. _____

11. ___ + ___ + ___ + ___ + ___ = ___

    ___ + ___ + ___ + ___ + ___ + ___ = ___

12. _____

13. 3^2 + 4^2 = ___^{2}

    3^3 + 4^3 + 5^3 = ___^{3}

14. ___ + ___ + ___ + ___ = ___

15. _____ ; _____

SET III EXERCISES

1. _____

2. _____

3. _____

4. _____ ; _____

Reinforcing Past Lessons

1. It has rained in a town for 14 consecutive weekends. Based on this record, a resident concludes that it will rain on every subsequent weekend. What kind of reasoning is being employed?

   _____

2. On a shoe rack (or "shoe tree") the number of slots for shoes in the tiers of the rack form an arithmetic sequence. The first tier has 2 slots and the sixth row has 22 slots.

   a. What is the common difference of this sequence?   _____

   b. Write in the missing terms of the sequence.   2   ___   ___   ___   ___   22

Name _____

Date _____

SET I EXERCISES

1.

t_1	t_2	t_3	t_4	t_5	t_6	t_7	t_8	t_9	t_{10}	t_{11}	t_{12}	t_{13}	t_{14}	t_{15}
1	1	2	3	5	8	13	21	__	__	__	__	__	__	__

2. _____

3. _____

4. _____ 5. _____

6. _____ _____ _____

7. _____

8. _____ _____

9. _____

10. _____

11. _____

12. _____ 13. _____ 14. _____

15. _____ 16. _____ 17. _____

SET II EXERCISES

1.   *Pattern A*

1	+	1											=	2
1	+	1	+	2									=	4
1	+	1	+	2	+	3							=	__
1	+	1	+	2	+	3	+	5					=	__
1	+	1	+	2	+	3	+	5	+	8			=	__
1	+	1	+	2	+	3	+	5	+	8	+	13		=	__
1	+	1	+	2	+	3	+	5	+	8	+	13	+	21 =	__

2. _____

3. _____

4. *Pattern B*

1^2 + 1^2 = 2

1^2 + 2^2 = 5

2^2 + 3^2 = ___

3^2 + 5^2 = ___

5^2 + 8^2 = ___

8^2 + 13^2 = ___

5. _____

6. _____

7. *Pattern C*

1^2 + 1^2 = 2 = 1 · 2

1^2 + 1^2 + 2^2 = 6 = 2 · 3

1^2 + 1^2 + 2^2 + 3^2 = 15 = 3 · 5

1^2 + 1^2 + 2^2 + 3^2 + 5^2 = 40 = 5 · 8

___ + ___ + ___ + ___ + ___ + ___ = ___ = ___ · ___

8. _____

9. *Pattern D*

1^3 + 2^3 – 1^3 = 8

2^3 + 3^3 – 1^3 = 34

3^3 + 5^3 – 2^3 = ___

5^3 + 8^3 – 3^3 = ___

10. _____

11. _____

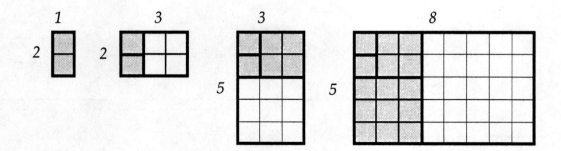

12. Use the graph space below to draw the next figure in the pattern.

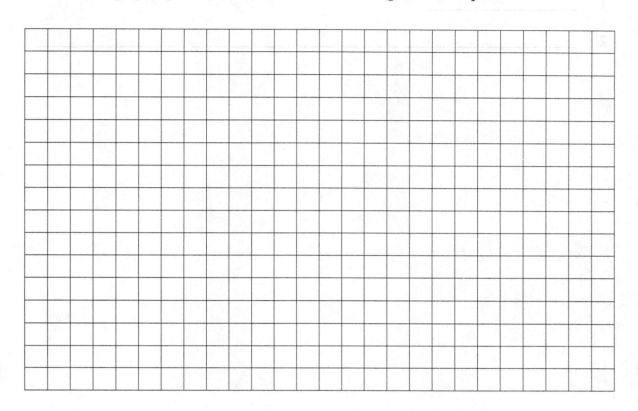

13.   8   13   21   34   55   89   144   233   377   610   987
       ∨   ∨   ∨   ∨   ∨   ∨   ∨   ∨   ∨   ∨

      ___   ___   ___   ___   ___   ___   ___   ___   ___   ___

14. _____

15. _____ ; _____

Using the Calculator

1. Using your calculator, write the next eight terms in the Fibonacci sequence given in Exercise 13 in Set II.

     987   ____   ____   ____   ____   ____   ____   ____   ____

2. Which (if any) of the terms listed in the sequence shown in the above question have repeated digits?

     _____

SET III EXERCISES

1. _____

2. _____

Summary and Review

Name _____

Date _____

SET I EXERCISES

1. _____   2. _____

3.   2   6   10   _____   18

4.   9   16   25   _____   49

5.   4   12   36   _____   324

6.   1   8   27   _____   125

7.   8   13   21   _____   55

8.   1   3   6   _____   15

9. _____   10. _____

11. _____

12. _____

13. _____

14.   1   =   1

    1 + 2 + 1   = _____

    1 + 2 + 3 + 2 + 1   = _____

    1 + 2 + 3 + 4 + 3 + 2 + 1   = _____

    1 + 2 + 3 + 4 + 5 + 4 + 3 + 2 + 1 = _____

15. _____

16. _____

SET II EXERCISES

1. $1{,}729 = \underline{\quad}^3 + \underline{\quad}^3$

    $1{,}729 = \underline{\quad}^3 + \underline{\quad}^3$

2. $50 = \underline{\quad}^2 + \underline{\quad}^2$

    $50 = \underline{\quad}^2 + \underline{\quad}^2$

3. _____ 4. _____ 5. _____

6.   5   17   _____

7. _____

8. 1 ____ 9 ____ ____ ____

9. 1 ____ 9 ____ ____ ____

10. 1 ____ 9 ____ ____ ____

11. _____ 12. _____

13. 5^2 = __   3 · 8 = __
    8^2 = __   5 · 13 = __
   13^2 = __   8 · 21 = __
   21^2 = __   13 · 34 = __

14. _____

15. __ = __   __ · __ = __

16. 1^2 + 2^2 + 2^2 = $\_\_^2$
   2^2 + 3^2 + 6^2 = $\_\_^2$
   3^2 + 4^2 + 12^2 = $\_\_^2$

17. __ + __ + __ = ____

18. _____ = ____

19. 10^2 + __ + __ = ____

20. _____ = ____

SET III EXERCISES

1. _____

2. _____ 3. _____

4. _____

5. _____ ; _____

Name _____

Date _____

SET I EXERCISES

The textbook defines a *function* as a pairing of two sets of numbers so that to each number in the first set there corresponds exactly one number in the second set. The definition begins with a pairing of two sets of numbers, called an "ordered pair." The x values are usually the first numbers in an ordered pair, and the y values are usually the second. Here is another way to think of the definition:

For every x there is exactly one y.

Two ways to represent functions are by *formulas* and *tables*.

1. $y = x + 4$

x	0	1	2	3	4
y	4	___	___	___	___

2. $y = 7x$

x	0	1	2	3	4
y	___	7	___	___	___

3. $y = 8 - x$

x	0	1	2	3	4
y	8	___	___	___	___

4. $y = \frac{12}{x}$

x	1	2	3	4	5
y	12	___	___	___	___

(Why is there no zero in this table? Think about what happens when you divide by zero.)

5. $y = 11x + 1$

x	1	2	3	4	5
y	12	___	___	___	___

6. $y = 2(x - 5)$

x	5	6	7	8	9
y	0	___	___	___	___

(Remember the order of operations—do "$x - 5$" first, then multiply by 2.)

7. $y = 0x + 3$

x	1	2	3	4	5
y	3	___	___	___	___

8. $y = x^2$

x	1	2	3	4	5
y	___	___	___	___	___

9. $y = x^2 + 2x + 1$

x	1	2	3	4	5
y	4	___	___	___	___

10. $y = (x - 1)^3$

x	2	3	4	5	6
y	__	__	__	__	__

11. $y = 2^x$

x	2	3	4	5	6
y	4	8	__	__	__

12. _____   13. _____

14. _____   15. _____

16. _____

Using the Calculator

Use the graphing calculator to graph selected problems from Set I. For the range for each problem, use the standard screen (x-values from -10 to 10, and y-values from -10 to 10) unless otherwise noted. Use the "trace" function to locate the points indicated for each graph. (Round to the nearest tenth.) Make a separate grid for each problem. Copy the graphs onto the graph paper provided by plotting the points that you located. Label each graph with the equation and unit scales for each variable.

1. $y = 7x$

x	-1	0	1
y	__	__	__

2. $y = \frac{12}{x}$

x	-9	-5	-1	1	5	9
y	__	__	__	__	__	__

3. $y = 0x + 3$

x	-5	0	5
y	__	__	__

a.   Clearly describe the graph for number 3.

_____

4. $y = x^2$

x	-3	-2	0	2	3
y	__	__	__	__	__

(To use an exponent, look for the "x^2" button, or the "^" button that indicates an exponent or power. You will need to use the "^" button for the 3rd power in number 5 and the "x" power in number 6.)

a.   Would a straight edge connect the points of the graph for number 4? ____

5. $y = (x - 1)^3$

x	-1	0	1	2	3
y	__	__	__	__	__

6. $y = 2^x$

x	−7	−3	0	2	3
y	___	___	___	___	___

SET II EXERCISES

1. $y = $ _____

x	3	4	5	6	7
y	6	8	10	12	14

2. $y = $ _____

x	2	3	4	5	6
y	10	11	12	13	14

3. $y = $ _____

x	7	8	9	10	11
y	4	5	6	7	8

4. $y = $ _____

x	3	4	5	6	7
y	9	16	25	36	49

5. $y = $ _____

x	2	3	4	5	6
y	22	33	44	55	66

6. $y = $ _____

x	5	6	7	8	9
y	51	61	71	81	91

7. $y = $ _____

x	6	7	8	9	10
y	28	33	38	43	48

8. $y = $ _____

x	1	2	3	4	5
y	1	8	27	64	125

9. $y = $ _____

x	1	2	3	4	5
y	2	9	28	65	126

10. $y = $ _____

x	1	2	3	4	5
y	2	6	12	20	30

11. | Depth of water in feet, x | 10 | 20 | 30 | 40 | 50 |
|---|---|---|---|---|---|
| Length of line in feet, y | 70 | 140 | 210 | 280 | ___ |

12. _____

13. _____

14. _____

15. | Percent of lifetime in hibernation, x | 0 | 10 | 20 | 30 |
|---|---|---|---|---|
| Expected lifespan in days, y | ___ | ___ | ___ | ___ |

16. _____

17. _____

18. | Age in years, x | 20 | 30 | 40 | 50 | 60 |
|---|---|---|---|---|---|
| Heart rate in beats per minute, y | 200 | 190 | 180 | 170 | ___ |

19. _____

20. _____

21. _____

22. Number of cylinders, *x*   4   6   8

Number of miles per gallon, *y*   ___   ___   ___

23. _____

SET III EXERCISES

| *x* | 1 | 4 | ___ | 16 | 25 |
| *y* | 0.5 | 1 | 1.5 | 2 | ___ |

1. _____   2. _____

3. _____

4. _____ ; _____

CHAPTER **3**

● **LESSON 2**

Name _____

Date _____

SET I EXERCISES

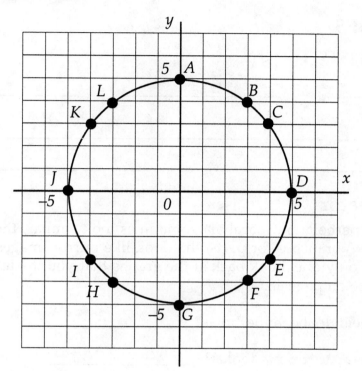

1. (4, 3) _____

2. (3, 4) _____

3. (−3, −4) _____

4. (4, −3) _____

5. (−4, 3) _____

6. (5, 0) _____

7. (−5, 0) _____

8. (0, 5) _____

9. Same x-coordinate as C _____

10. Same y-coordinate as H _____

11. A _____   B _____   C _____   D _____   E _____   F _____

12. A _____   B _____   C _____   D _____   E _____   F _____   G _____

13. Complete on the graph paper provided.

SET II EXERCISES

1–6. Plot graphs on the paper provided.

7. _____   8. _____   9. _____

10. _____   11. _____   12. _____

13. _____ 14. _____ 15. _____

16. _____

SET III EXERCISES

1. Plot graph on the paper provided.

2. _____ 3. _____ 4. _____

5. _____

_____

Using the Calculator

1. Although the hyperbola and parabola sometimes look similar, the hyperbola has two branches (or parts, that look like mirror images) while the parabola has only one. Look back at the graphs that you made in the calculator exercise on page 48.

   a. Which ones appear to be hyperbolas? _____

   b. Which ones appear to be parabolas? _____

2. Graph each of the following functions using your graphing calculator (or you can determine the values for y and plot instead). To graph, use the "trace" function and determine the values of y to the nearest whole number. For range, use the standard values of x (−10, 10) and y (−10, 10).

   a. $y = \frac{1}{x}$

x	−8	−5	−3	−1	1	3	5	8
y								

   b. What happens at $x = 0$? _____

   c. $y = \frac{-1}{x}$

x	−8	−5	−3	−1	1	3	5	8
y								

   (Be sure to use the "−" that stands for "negative" and not "subtraction.")

   d. What curves do they appear to be? _____

   e. Compare both graphs. How are they different or alike?

   _____

   _____

SET I EXERCISES

1. $y = 2x + 3$

x	0	1	2	3	4
y	___	___	___	___	___

2. _____

3. Graph on the paper provided.

4.
x	2	3	4	5	6
y	6	7	8	9	10

   $y =$ _____

5. Graph on the paper provided.

6. _____   7. _____

8.
x	3	4	5	6	7
y	1	___	___	___	___

9. _____

10. Function A: $y = 2x + 1$

x	0	1	2	3	4
y	___	___	___	___	___

    Function B: $y = 3x + 1$

x	0	1	2	3	4
y	___	___	___	___	___

    Function C: $y = 4x + 1$

x	0	1	2	3	4
y	___	___	___	___	___

11. Graph on the paper provided.

12. _____   13. _____   14. _____

15. _____

16. Function D: $y = x + 3$

x	0	1	2	3	4
y	___	___	___	___	___

    Function E: $y = x + 5$

x	0	1	2	3	4
y	___	___	___	___	___

Function F: $y = x + 8$

x	0	1	2	3	4
y	___	___	___	___	___

17. Graph on the paper provided.   18. _____

19. _____   20. _____

21. Function G: $y = 6 + x$

x	0	1	2	3	4
y	___	___	___	___	___

Function H: $y = 6 - x$

x	0	1	2	3	4
y	___	___	___	___	___

22. Graph on the paper provided.   23. _____

SET II EXERCISES

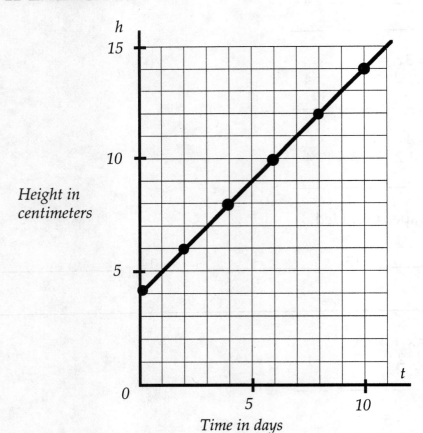

Height in centimeters

Time in days

1.

t	0	2	4	6	8	10
h	___	___	___	___	___	___

2. _____   3. _____

4. $h = 10 - 2t$

t	0	1	2	3	4	5
h	__	__	__	__	__	__

5. Graph on the paper provided.

6. _____ 7. _____ 8. _____

9. _____ 10. _____

11. _____

12. Graph on the paper provided.   13. _____

14. $y = 6x - 40$

x	50	60	70	80	90
y	260	__	__	__	__

15. _____

16. _____   17. Graph on the paper provided.

18. _____

19. Graph on the paper provided.   20. _____

21. _____

Using the Calculator

Check your equation in Exercise 13 by entering the equation in the graphing calculator. Use x-range of $(-20, 20)$ and y-range of $(-4, 250)$. Trace the x-values from the table that was given. Tracing will give approximate values only, but you can see if your equation is correct. Record each y-value as you trace, rounding to three decimal places (thousandths).

x	8	10	12	14
y	__	__	__	__

SET III EXERCISES

1. $y = 263 - 0.13x$ _____

2.

x	1900	1930	1970	1980
y	16	____	____	____

3. Graph on the paper provided.   4. _____

5. _____

Reinforcing Past Lessons

1. Given the following set of x and y values, determine the equation of each function.

a.

x	0	1	2	3	4	5
y	2	3	6	11	18	27

Equation _____

b.

x	0	1	2	3	4	5	6
y	3	5	19	57	131	253	435

Equation _____

c.

x	0	1	2	3	4	5
y	1	4	7	10	13	16

Equation _____

d.

x	0	1	2	3	4	5
y	−4	−3	0	5	12	21

Equation _____

e.

x	0	1	2	3	4	5
y	2	5	26	83	194	377

Equation _____

SET I EXERCISES

1. | x | 0 | 1 | 2 | 3 | 4 |
    |---|---|---|---|---|---|
    | y | 0 | 1 | 4 | 9 | 16 |

    $y = $ _____

2. Graph on the paper provided.

3. | x | −4 | −3 | −2 | −1 |
    |---|---|---|---|---|
    | y | 16 | ___ | ___ | ___ |

4. Function A: $y = x^2 + 2$

x	−3	−2	−1	0	1	2	3
y	—	—	—	—	—	—	—

    Function B: $y = x^2 + 10$

x	−3	−2	−1	0	1	2	3
y	—	—	—	—	—	—	—

5. Graph on the paper provided.

6. _____

7. _____

8. _____

9. $y = 12 - x^2$

x	−3	−2	−1	0	1	2	3
y	—	—	—	—	—	—	—

10. Graph on the paper provided.

11. _____

12. _____

13. _____

14. $y = (3 - x)^2$

x	0	1	2	3	4	5	6
y	—	—	—	—	—	—	—

15 Graph on the paper provided.

16. _____

17. _____

18. _____

SET II EXERCISES

1. $I = 2s^2$

s	0	1	2	3	4
I	—	—	—	—	—

2. _____

3. _____

4.   Graph on the paper provided.   5.   Graph on the paper provided.

6.   _____

7.
x	0	10	20	30	40	50	60
y	___	___	160	___	___	___	___

8.   Graph on the paper provided.   9.   _____

10. _____   11. _____

12. $y = 88 - 0.12x^2$
x	0	5	10	15	20	25	27
y	___	85	___	___	___	___	___

13. Graph on the paper provided.   14. _____

15. _____   16. _____

SET III EXERCISES

1. $y = 0.07x^2 - x + 8.25$
x	10	12	15	18
y	___	___	___	___

2.   Graph on the paper provided.   3.   _____   4.   _____

5.   _____   6.   _____

Using the Calculator

1.   Use your calculator to graph the equations given. Fill in the table for the y-values to the nearest hundredth (two decimal places). The "standard seven" values for x are used. Graph on the paper provided.

a.   Standard range: (x: –10, 10), (y: –10, 10)
$y = x^2 - 7$
x	–3	–2	–1	0	1	2	3
y	___	___	___	___	___	___	___

b.   Range: (x: –10, 10), (y: –15, 10)
$y = 2x^2 - 12$
x	–3	–2	–1	0	1	2	3
y	___	___	___	___	___	___	___

2.   Use a graphing calculator to draw the graph of this function. Use "trace" to complete the table to the nearest hundredth. Graph these points on the paper provided. Label your equation and indicate the scales for x and y.

a.   Use range of (x: –20, 20), (y: –40, 40)
$y = 35 - .46x^2$
x	–10	–8	–4	0	4	8	10
y	___	___	___	___	___	___	___

SET I EXERCISES

1. $y = \frac{6}{x}$

x	1	2	3	4	5	6
y	—	—	—	1.5	—	—

2. Graph on the paper provided.

3. _____

4. _____

5. _____

6. $y = \frac{6}{x}$

x	−6	−5	−4	−3	−2	−1
y	−1	−1.2	—	—	—	—

7.

x	−4	−2	−1	1	2	4
y	−1	—	—	—	—	—

8. _____

9. $y = x^3$

x	−4	−3	−2	−1	0	1	2	3	4
y	−64	—	—	—	—	—	—	—	—

10. Graph on the paper provided.

11. _____

12. _____

13. _____

14. _____

15. _____

16. _____

Using the Calculator

1. Graph the following equations with the graphing calculator, using the ranges specified. Trace to find the values for y and round to the nearest hundredth to complete the tables. Graph on the paper provided. Be sure to indicate the scales and the equations.

a. Range: $(x: -4, 4)$, $(y: -2, 20)$

$y = x^4 + 2$

x	−2	−1	0	1	2
y	—	—	—	—	—

b.   Range: $(x: -3, 3)$, $(y: -40, 40)$, $y-$ scale $= 5$

$y = x^5 + 3$	x	-2	-1	0	1	2
	y	—	—	—	—	—

SET II EXERCISES

1. _____   2. _____

3. $y = \frac{330}{x}$

x	20	40	60	80	100	150
y	—	—	—	—	—	—

4. _____   5.   Graph on the paper provided.

6. $v = 2.2l^3$

l	1	5	8	10	12	15
v	2.2	—	—	—	—	—

7. _____   8. _____

9. _____   10. _____

11.   Graph on the paper provided.

12. $y = \frac{57,600}{x^2}$

x	15	30	60	120	240
y	—	—	—	—	—

13. _____

14.   Graph on the paper provided.

SET III EXERCISES

1. _____   2. _____

3.   Graph on the paper provided.   4. _____

Reinforcing Past Lessons

1. How much do you remember about number sequences from the material in Chapter 2? Exercises a–k recall information contained in Lessons 1–6 of that chapter.

a. Identify the sequence below as either arithmetic or geometric.

   3 15 75 375 _____

b. What is the common difference or common ratio of the sequence in the above question? _____

c. Find the 13th term of the sequence below.

   3 10 17 24 31 38 ... _____

d. How did you arrive at your answer to the above question? Explain your calculation.

   _____

e. Determine the decimal forms of the following binary numerals.

   11001 _____   110 _____

   1010 _____   11111 _____

f. Determine the binary forms of the following decimal numerals.

   27 _____   33 _____

   49 _____   104 _____

g. Give the 5th through 9th terms in the sequence of squares.

   _____

h. Find the digital root of the number 345,891. _____

i. There are two ways to complete this number to make it a palindrome: 24698. What are they?

   24698_____ and 24698_____

j.   List the 4th through 8th terms of the sequence of cubes.

_____

k.   List the 5th through 9th terms of the Fibonacci sequence.

_____

The questions in Exercise 2 relate to material that you learned earlier in this chapter.

2.   The "Golden Ratio" is a special measurement for a rectangle that is pleasing to the eye. It can be seen everywhere from the Parthenon of Athens to a television screen. Notebook paper (usually 8-1/2" by 11") and photo sizes (often 3-1/2" by 5", 8" by 10", or 11" by 14") also exhibit this ratio.

a.   For the purposes of this exercise, let the function be: $y = 1.3x$. Determine the values for y if we consider some common measures for photos and frames.

x	2	3.5	8	11	14
y	—	—	—	—	—

b.   How do the values that you determined compare with those that are used in photo sizes?

_____

_____

Name _____

Date _____

SET I EXERCISES

To *interpolate* is to guess values of a variable *between* values that are known.

To *extrapolate* is to guess values of a variable *beyond* those that are known.

1. Graph on the paper provided.

2. _____   3. _____

4. _____ ; _____

5. _____   6. _____ ; _____

7. _____

8. Graph on the paper provided.   9. _____

10. _____ 11. _____ 12. _____

13. _____

SET II EXERCISES

1. Graph on the paper provided.

2. _____ 3. _____ 4. _____

5. _____ ; _____

6. Graph on the paper provided.
(A "smooth curve" may not need the use of a straight edge.)

7. _____

8. _____ 9. _____ 10. _____

11. _____

12. Graph on the paper provided.   13. _____

14. _____

15. _____

Supplemental Exercises

1. The table below gives the number of never married persons as a percentage of the total U. S. population by gender from the years 1970–1990.

	1970	1980	1985	1990
Male	18.9	23.8	25.2	26.1
Female	13.7	17.1	18.2	19.3

a. Complete the graph below using both sets of statistics. Use different colors for male and female. Draw a line through the values, and continue the line to complete the graph.

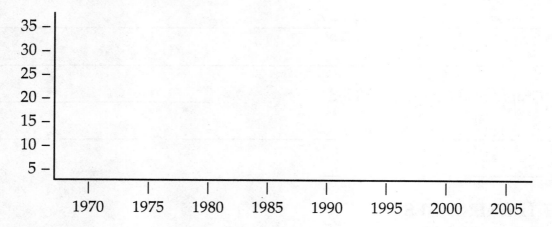

b. Guess the percent in 1975 for males _____, and females _____ .
Did you *interpolate* or *extrapolate* to get your answer? _____

c. Guess the percent in 2000 for males _____, and females _____ .
Did you *interpolate* or *extrapolate* to get your answer? _____

SET III EXERCISES

1. _____

2. _____   3. Graph on the paper provided.

4. 1970: _____   1980: _____   1990: _____

5. _____   6. _____

7 _____

8. _____

Supplemental Exercises

1.   This table gives the U.S. population in the years indicated, in millions to the nearest one hundred thousand. Complete the graph below using both sets of statistics. Connect the points with a smooth curve to the year 2030.

Year, x	1790	1830	1870	1910	1950	1990
Population, y	3.9	12.9	39.8	92.0	151.3	248.7

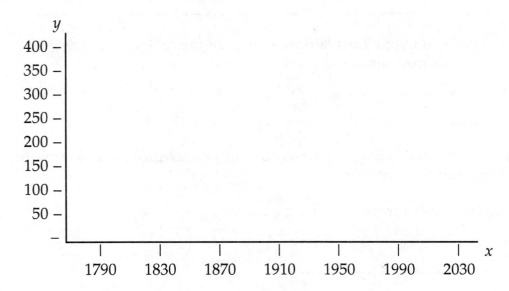

b.   Guess the the population for the year 2030 _____

Did you *interpolate* or *extrapolate* to get your answer? _____

Using the Calculator

Use your graphing experience from this chapter to graph the *general grouping* in each set below. After you draw what you believe each group should look like, check with a graphing calculator or try some values to see how the graph should appear. Draw on the axes given.

1. $y = x^3, y = x^5, y = x^7$ 2. $y = x^2, y = x^4, y = x^6$

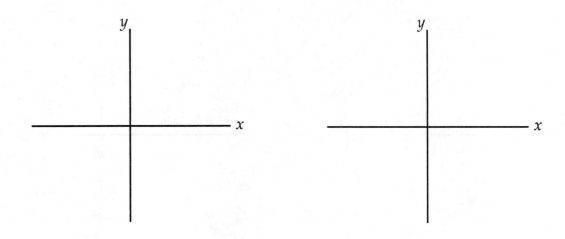

3. Use the graphing calculator to graph the following equation: $y = \frac{5}{x}$
   Use the standared screen with x– range (–10, 10) and y– range (–10, 10).

4. What happpens at $x = 0$?

   _____

5. What is your explanation for what happens at $x = 0$?

   _____

6. Trace the curve on your calculator to complete the table. Round the y-values to the nearest tenth.

x	–8	–4	–1	0	1	4	8
y	___	___	___	___	___	___	___

7. Graph the equation in Exercise 4 on the graph paper provided. Be sure to label scales and the equation.

8. Complete the table for the following equation: $y = x^3 - 4$

x	0	1	2	3	4	5
y	___	___	___	___	___	___

SET I EXERCISES

1. A = __   2. B = __   3. C = __   4. D = __

5.

x	80	90	100	110
y	50	____	____	____

6. $y =$ _____

7. _____   8. _____

9. _____   10. _____

11. Function A: $y =$ _____

x	0	1	2	3	4	5
y	0	4	8	12	16	20

(Use repeated subtraction to help.)

12. Function B: $y =$ _____

x	0	1	2	3	4	5
y	0	1	4	9	16	25

13. Graph on the paper provided.

14. _____   15. _____

Supplemental Exercises

1. Complete the tables for each function.

a. $y = 3x^2 + 5$

x	0	1	2	3	4	5
y	___	___	___	___	___	___

b. $y = x^3 - 1$

x	0	1	2	3	4	5
y	___	___	___	___	___	___

2. Find the equation for each set of x and y values using repeated subtraction.

a. $y =$ _____

x	0	1	2	3	4	5
y	4	6	20	58	132	254

b. $y =$ _____

x	0	1	2	3	4	5
y	1	13	25	37	49	61

SET II EXERCISES

1. $y = \dfrac{120}{x}$

x	1	2	3	4	5	6
y	___	___	___	___	___	___

2. _____

3. _____

4. Graph on the paper provided.

5. $y = 0.6x^2$

x	0	10	20	30	40	50
y	___	60	___	___	___	___

6. Graph on the paper provided.   7. _____

8. _____

9. Graph on the paper provided.

10. _____   11. _____

12. _____   13. _____

14. _____ ; _____

_____

SET III EXERCISES

1. $w = \dfrac{2560}{(d+4)^2}$

d	0	1	2	3	4	5	6
w	___	___	___	___	___	___	___

2. _____

3. _____

4. $w = 160 + 40d$

d	0	−1	−2	−3	−4
w	___	___	___	___	___

5. _____   6. Graph on the paper provided.

7. _____

Name _____

Date _____

SET I EXERCISES

A number is in *scientific notation* if it is written in the form

$$a \times 10^b$$

in which *a* is a number that is at least 1 but less than 10. The number *a* is called the *coefficient* and the number *b* is called the *exponent*.

1. _____ 2. _____

3. _____ 4. _____

5. _____

6. Coefficient _____ Exponent _____

7. _____ 8. _____

$$\boxed{635 \times 10^9}$$

9. _____

10. _____ 11. _____

12. _____ 13. _____

Color of hair	Number of hairs
Black or Brown	1.05×10^5
Blond	1.4×10^5
Red	9×10^4

14. _____

15. _____

_____

$$\boxed{\text{The radius of Antares is } 2 \times 10^8 \text{ kilometers.}}$$

16. _____

17. Radius _____ Diameter _____

$$\boxed{2,000,744}$$

18. _____   19. _____

20. _____

Supplemental Exercises

1. In 1950, the population of the world was about 2.516 billion. Write this number in decimal form. _____

2. In 1990, the population of the world was about 6.261 billion. Write this number in decimal form. _____

3. Liechtenstein, one of the world's smallest countries, had a population of about 29,000 in 1993. Write this number in scientific notation. _____

SET II EXERCISES

$$\boxed{\text{Each square centimeter of skin has about } 4 \times 10^6 \text{ microbes.}}$$

$$\boxed{\text{Each person has about } 2 \times 10^4 \text{ square centimeters of skin.}}$$

1. _____   2. _____

3. _____   4. _____

5. $(2 \times 10^{10}) \times (3 \times 10^5) =$ _____

6. $(5 \times 10^8) \times (1.6 \times 10^{12}) =$ _____

$$
\begin{aligned}
(3 \times 10^4) \times (7 \times 10^2) &= 21 \times 10^6 \\
&= 2.1 \times 10 \times 10^6 \\
&= 2.1 \times 10^7
\end{aligned}
$$

7. $(5 \times 10^3) \times (6 \times 10^8) =$ _____

8. $(8 \times 10^{10}) \times (7.5 \times 10^5) =$ _____

9. $(9 \times 10^9) \times (9 \times 10^9) =$ _____

10. 500,000 _____   26,000,000 _____

11. _____

12. _____

13. _____

14. $\dfrac{750,000}{50,000}$ = [answer to 13] _____

15. _____
_____

16. $\dfrac{8 \times 10^{10}}{2 \times 10^{10}}$ _____

17. $\dfrac{9 \times 10^{15}}{5 \times 10^{5}}$ _____

18. $\dfrac{3 \times 10^{40}}{6 \times 10^{5}}$ _____

Mammal	Number of neurons
Human	3×10^{10}
Gorilla	7.5×10^{9}
Cat	6.5×10^{7}

19. _____

20. _____

21. _____

22. $\dfrac{9.3 \times 10^{7}}{1.86 \times 10^{5}}$ _____

23. _____

Sound travels a distance of about 7.7×10^2 miles each hour through air.

The sun is about 93,000,000 miles from the earth.

24. _____

SET III EXERCISES
Show your work.

1. _____

2. _____

3. _____

4. _____

5. _____

Reinforcing Past Lessons
Without looking back at Lesson 1, write each of the following in words.

1. $3 \times 10^{30} =$ _____

2. $17 \times 10^{24} =$ _____

3. $132 \times 10^{12} =$ _____

4. $435 \times 10^{33} =$ _____

5. $50 \times 10^{21} =$ _____

6. $194 \times 10^{15} =$ _____

Name _____

Date _____

SET I EXERCISES

Number	Log	Number	Log	Number	Log
1	0	2,048	11	4,194,304	22
2	1	4,096	12	8,388,608	23
4	2	8,192	13	16,777,216	24
8	3	16,384	14	33,554,432	25
16	4	32,768	15	67,108,864	26
32	5	65,536	16	134,217,728	27
64	6	131,072	17	268,435,456	28
128	7	262,144	18	536,870,912	29
256	8	524,288	19	1,073,741,824	30
512	9	1,048,576	20	2,147,483,648	31
1,024	10	2,097,152	21	4,294,967,296	32

1. 128
    × 32

2. 128 $\longrightarrow$
    32 $\longrightarrow$ +
    $\longleftarrow$

3. 1024
    × 512

4. 1024 $\longrightarrow$
    512 $\longrightarrow$ +
    $\longleftarrow$

5. 65536 $\longrightarrow$
    16 $\longrightarrow$ +
    $\longleftarrow$

6. 16384 $\longrightarrow$
    8192 $\longrightarrow$ +
    $\longleftarrow$

7.   2048 $\longrightarrow$ _____
       256 $\longrightarrow$
        64 $\longrightarrow$ _____
             $\longleftarrow$

8.   4096 $\longrightarrow$ 12
       128 $\longrightarrow$ 7
             $\longleftarrow$

9. _____

10.
    $2048 \overline{)32768}$

11.   32768 $\longrightarrow$
      $\div$ 2048 $\longrightarrow$ _____
                  $\longleftarrow$

12.   262144 $\longrightarrow$
      $\div$ 256 $\longrightarrow$ _____
                 $\longleftarrow$

13.   16384 $\longrightarrow$
      $\div$ 16384 $\longrightarrow$ _____
                   $\longleftarrow$

14.   536870912 $\longrightarrow$
      $\div$ 64 $\longrightarrow$ _____
                    $\longleftarrow$

15. _____

16. _____

17. $4^5 =$

18. $4^5 =$

19. $128^2 =$

20. $64^3 =$

21. $32^4 =$

22. $8^{10} =$

● SET II EXERCISES

Number		Log
1	=	0
2	= 2	1
4	= 2×2	2
8	= $2 \times 2 \times 2$	3
16	= $2 \times 2 \times 2 \times 2$	4
32	= $2 \times 2 \times 2 \times 2 \times 2$	5
64	= $2 \times 2 \times 2 \times 2 \times 2 \times 2$	6
128	= $2 \times 2 \times 2 \times 2 \times 2 \times 2 \times 2$	7
256	= $2 \times 2 \times 2 \times 2 \times 2 \times 2 \times 2 \times 2$	8
512	= $2 \times 2 \times 2 \times 2 \times 2 \times 2 \times 2 \times 2 \times 2$	9
1,024	= $2 \times 2 \times 2 \times 2 \times 2 \times 2 \times 2 \times 2 \times 2 \times 2$	10
2,048	= $2 \times 2 \times 2 \times 2 \times 2 \times 2 \times 2 \times 2 \times 2 \times 2 \times 2$	11
4,096	= $2 \times 2 \times 2 \times 2 \times 2 \times 2 \times 2 \times 2 \times 2 \times 2 \times 2 \times 2$	12

● 1. _____   2. _____

3. _____   4. _____

5.

6.

7.

●

Number			Log
1	=	2^0	0
2	=	2^1	1
4	=	2^2	2
8	=	2^3	3
16	=	2^4	4
32	=	2^5	5
64	=	2^6	6
128	=	2^7	7
256	=	2^8	8
512	=	2^9	9
1,024	=	2^{10}	10
2,048	=	2^{11}	11
4,096	=	2^{12}	12

8. _____   9. _____

10. _____   11. _____

12. $32 \times 128 = 4{,}096$ _____

13. $32 =$ _____ $128 =$ _____ $4{,}096 =$ _____

14. _____

15. $2{,}048 \div 8 = 256$ _____

16. $2{,}048 =$ _____ $8 =$ _____ $256 =$ _____

17. _____

18. $16^3 = 4{,}096$ _____

19. $16 =$ _____ $4{,}096 =$ _____

20. _____

SET III EXERCISES

1. _____   2. _____

3. _____   4. _____

CHAPTER **4**

LESSON 4

SET I EXERCISES

Numbers	1	2	3	4	5	6	7	8	9	10
Logarithms	0	0.30	0.48	0.60	0.70	0.78	0.85	0.90	0.95	1

1. Log 2 + log 6 = log 12 _____

2. _____   3. _____

4. _____   5. Log 14 = _____

6. Log 15 = _____   7. Log 16 = _____

8. Log 20 = _____

9.
Numbers	10	20	30	40	50	60	70	80	90	100
Logarithms	__	__	__	__	__	__	__	__	__	__

10.
Numbers	10	10^2	10^3	10^4	10^5	10^6	10^7	10^8	10^9	10^{10}
Logarithms	__	__	__	__	__	__	__	__	__	__

11. Log 10^{15} = _____

12. Log 1,000,000,000,000,000,000 = _____

13. Log 40 = _____   14. Log 10^{95} = _____

15. _____   16. _____   17. _____

Reinforcing Past Lessons

Can you recall names of the mathematical terms from earlier chapters?

1. Another name for base 10 numbers is _____.

2. Another name for base 2 numbers is _____.

3. Guessing at the values of a variable *between* those that are already known is called _____.

4. Guessing at the values of a variable *beyond* those that are already known is called _____.

5. The number that is added to each term in an arithmetic sequence to get the next term is called the _____.

6. The number that is multiplied by each term in an geometric sequence to get the next term is called the _____.

SET II EXERCISES

Numbers	1	2	3	4	5	6	7	8	9
Logarithms	0	0.30	0.48	0.60	0.70	0.78	0.85	0.90	0.95

Numbers	10	10^2	10^3	10^4	10^5	10^6	10^7	10^8	10^9
Logarithms	1	2	3	4	5	6	7	8	9

1. Log of 3×10^6 = _____

2. Log of 4×10^8 = _____

3. Log of 7×10^5 = _____

4. Log of 1×10^9 = _____

5. Log of 6×10^{12} = _____

6. _____ 7. _____ 8. _____

9. _____ 10. _____

11. _____ 12. _____

13. _____

	Logarithm	Scientific notation	Decimal form
14.	6.48	_____	_____
15.	2.85	_____	_____
16.	11.7	_____	_____

SET III EXERCISES

1. _____ 2. _____

3. _____

4. _____

Name _____

Date _____

SET I EXERCISES

	Number in decimal form	Number in scientific notation	Logarithm of number
	121,000	1.21×10^5	$0.083 + 5 = 5.083$
1.	1.44		_____
2.	3.65		_____
3.	8.2		_____
4.		1.44×10^3	_____
5.		3.65×10^8	_____
6.		8.2×10^{20}	_____
7.	22,000	_____	_____
8.	3,150,000,000,000	_____	_____
9.	45	_____	_____
10.	1,000,000,000	_____	_____
11.	_____		0.017
12.	_____		0.107
13.	_____		0.17
14.	_____	_____	5.017
15.	_____	_____	2.107
16.	_____	_____	10.17

	Number in decimal form	Number in scientific notation	Logarithm of number
17.	_____		0.021
18.	_____		0.210
19.	_____		2.100
20.	_____		21.000

21.

Number	1	2	3	4	5	6	7	8
Log	0	0.301	____	____	____	____	____	____

22. _____   23. _____

24. _____   25. _____

26. Log of 256 = _____   Log of 512 = _____

SET II EXERCISES

DECIBELS

A decibel is an arbitrary unit based on the faintest sound that a man can hear. The scale is logarithmic, so that an increase of 10 db means a tenfold increase in sound intensity; a 20-db rise a hundredfold increase, and 30 db a thousandfold increase.

1.

Decibels	Loudness	Log of loudness
0	1	0
10	10	1
20	100	2
30	1,000	3
40	____	____
50	____	____
60	____	____
70	____	____
80	____	____

2. _____

3. _____   4. _____   5. _____

6. _____

7. _____

Interval of time	Number of seconds	8. Scientific notation	9. Log
1 second	1	_____	____
1 minute	60	_____	____
1 hour	3,600	_____	____
1 day	86,000	_____	____
1 year	32,000,000	_____	____
1 century	3,200,000,000	_____	____

10.

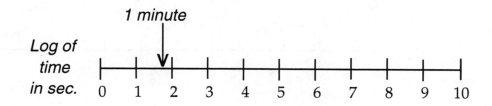

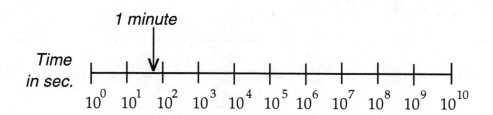

11. _____

Wave	Frequency in hertz	12. Scientific notation	13. Log
AC current	60	_____	_____
AM radio station at 980	980,000	_____	_____
TV channel 2	57,000,000	_____	_____
Microwave in oven	2,450,000,000	_____	_____
Red light	400,000,000,000,000	_____	_____
X-rays	30,000,000,000,000,000	_____	_____

14.

Log of frequency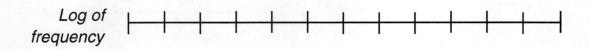

Frequency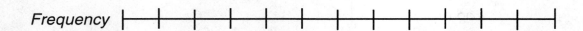

SET III EXERCISES

1. _____

2. _____   3. _____

CHAPTER 4
LESSON 6

Name _____

Date _____

SET I EXERCISES

1. _____

2.

Magnitude, x	0	1	2	3	4	5	6
Energy released, y	1	10	___	___	___	___	___

3. Graph on the paper provided.

4. _____

5. _____   6. _____

7. _____

8. _____   9. Graph on the paper provided.

10. _____   11. _____

12. Graph on the paper provided.

13.

f-stop, x	1	2	3	4	5	6	7	8	9
Log of f-stop number, y	0.15	0.30	___	___	___	___	___	___	___

14. Graph on the paper provided.   15. _____

Using the Calculator

1. Use the graphing calculator to graph the exponential curve with the equation $y = 4^x + 3$.

2. After entering the equation into the $y =$ ____ function, change the range to x: −2 to 4, x– scale = .3, and y: −20 to 100, y–scale = 10.

3. On graph paper, count by 1's on the x-axis and by 10's on the y-axis. In order to plot the graph accurately, use the trace button on the curve, and locate the points that complete the ordered pairs.

x	−1.7	−.5	0	.8	1.4	2.2	3.0	3.3
y	___	___	___	___	___	___	___	___

SET II EXERCISES

1. Graph on the paper provided.

Year	Millions of barrels	2. Logarithms
1880	30	_____
1890	77	_____
1900	150	_____
1910	330	_____
1920	690	_____
1930	1,400	_____
1940	2,200	_____
1950	3,800	_____
1960	7,700	_____
1970	17,000	_____

3. Graph on the paper provided.　　　4. _____

5. Graph on the paper provided.

6. Time in hours, x 　0　1　2　3　4　5　6
   Amount left in grams, y ___ ___ ___ ___ ___ ___ ___

7. Graph on the paper provided.　　　8. _____

9. _____　　10. _____

SET III EXERCISES

1. Graph on the paper provided.　　　2. _____

3.

Logarithms	Time until test in hours, x	Average score on test, y
_____	1	60
_____	2	53
_____	8	40
_____	24	30
_____	48	23
_____	96	17
_____	192	10

4. Graph on the paper provided.　　　5. _____

6. Log of 0.5 = _____　　log of 0.25 = _____

Summary and Review

Name _____

Date _____

SET I EXERCISES

1. _____

2. _____

3. _____

4. _____

5. _____

6. _____

7. _____

8. _____

9. _____

10. _____

11. _____

12. 729
    $\times$ 81

13. _____

14.

$9\overline{)2187}$

15. _____

SET II EXERCISES

Creature	Mass in miligrams	1. Scientific notation	2. Log
Spider	80	_____	____
Hummingbird	2,000	_____	____
Chicken	3,150,000	_____	____
Human	70,000,000	_____	____
Elephant	6,300,000,000	_____	____
Whale	138,000,000,000	_____	____

3.

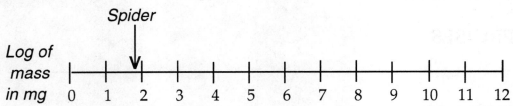

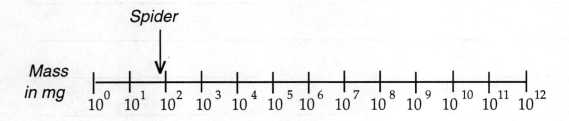

4. _____

5. _____

6. _____

7. Position of A, x 1   2   3   4   5   6   7   8

Position of A, x	1	2	3	4	5	6	7	8
Frequency, y	___	55	110	220	440	880	___	___

8. Graph on the paper provided.

9.

Position of A, x	1	2	3	4	5	6	7	8
Logarithm	___	___	___	___	___	___	___	___

10. _____   11. Graph on the paper provided.

12. _____

SET III EXERCISES

1. Graph on the paper provided.

2.

Depth in meters, x	0	50	100	150	200	250
Logarithm	___	___	___	___	___	___

3. Graph on the paper provided.   4. _____

5. _____

Name _____

Date _____

SET I EXERCISES

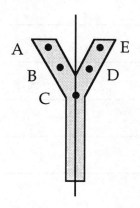
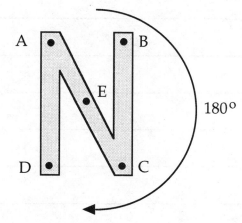

1. _____ 2. _____ 3. _____ 4. _____

5. _____

6. _____

7. _____ 8. _____ 9. _____

10. _____ 11. _____

12. _____ 13. _____

14. _____ ; _____

(Placing a mirror along the center of figures can be helpful in determining the symmetry of different figures. When you place the mirror on the line of symmetry, the reflection and the half of the figure that is visible should form a complete figure.)

15. _____

16. _____

17. _____

18. _____

19. _____ 20. _____ 21. _____

22. _____

(For line symmetry, note the number of lines. For rotational symmetry, include the number-fold of symmetry.)

23. _____

24. _____

25. _____

26. _____

27. _____

28. _____

29. _____

30. _____

Supplemental Exercises

1. Draw a rectangle in the space to the right.

    a. Does it have line symmetry? _____

    b. Draw the line(s) of symmetry, if any.

    c. Does it have rotational symmetry? _____

2. Consider the peace symbol.

    a. Does it have line symmetry? _____

    b. Draw the line(s) of symmetry, if any.

    c. Does it have rotational symmetry? _____

4. Consider these astrological symbols.

♊	♋	♌	♍	♎	♏
Gemini	Cancer	Leo	Virgo	Libra	Scorpio
♐	♑	♒	♓	♈	♉
Sagittarius	Capricorn	Aquarius	Pisces	Aries	Taurus

    a. Which, if any, have line symmetry? _____

    _____

    b. Which, if any, have rotational symmetry? _____

    _____

SET II EXERCISES

1. _____

2. _____

3. _____

4. _____

5. _____

6. _____

7. _____

8. _____

9. _____ 10. _____ 11. _____ 12. _____

13.
Figure	A	B	C	D
Number of sides, n	___	___	___	___
Angle of mirrors, m	120°	90°	___	___

14. _____

15. _____

16.
Number of sides, n	8	9	10	11
Angle of mirrors, m	120°	90°	___	___

SET III EXERCISES

1. _____
   _____

2. _____
   _____

3. _____

4. _____

Supplemental Exercises

Cut out a snowflake from a sheet of paper and look for rotational symmetry. To make a snowflake, take a square sheet of white paper, fold the paper in half, and in half again from the center, making sure that each fold is centered on the square. Cut a design on each line of the fold. Open up the folds and examine.

1. Is there rotational symmetry? If so, what type?

    _____ ; _____

2. Draw all lines of symmetry on the snowflake. How many lines are there?

    _____

3. Consider the symbols below.

    A.   B.   C.   D.

    $ ¥   %   §

    E.   F.   G.   H.

    ♂   ♀   🚗   🚗

    I.   J.   K.

    ☠   ♻   ⚛

    a. Which, if any, have line symmetry?   _____

    b. Which, if any, have rotational symmetry?   _____

Reinforcing Past Lessons

1. In 1989 the population of the world was 5,201,000,000. Write this number in scientific notation.

    _____

2. World production of wheat in 1989 was 538,000,000 short tons. Write this number in scientific notation.

    _____

3. In 1970 U.S. production of coal was 613,000,000 short tons. In 1989 U.S. coal production had grown to 1,036,000,000 short tons. Write both figures in scientific notation.

    _____   _____

Name _____

Date _____

SET I EXERCISES

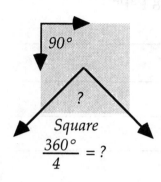

Square
$$\frac{360°}{4} = ?$$

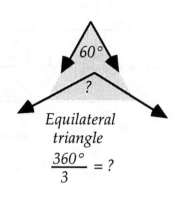

Equilateral
triangle
$$\frac{360°}{3} = ?$$

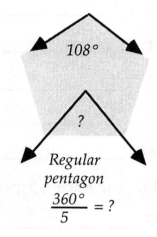

Regular
pentagon
$$\frac{360°}{5} = ?$$

1. _____

2.

Number of sides, n	3	4	5
Mirror angle, m	___	___	___
Angles of polygon	60°	90°	108°

3. _____ 4. _____

5. _____

6. _____ 7. _____ 8. _____

9. _____

10. _____ 11. _____

12. 13.

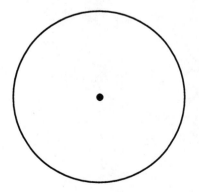 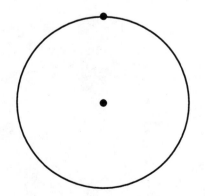

SET II EXERCISES

1. _____

2. _____   3.   Use tracing paper.

4. _____

5. _____   6.   _____

7. _____

8. _____   9.   _____

10. Use tracing paper.   11.   _____

12. _____

13. Use the figure from Exercise 10.   14.   _____

15. _____   16.   Use tracing paper.   17. _____

18. _____   19.   _____

SET III EXERCISES

Name _____

Date _____

SET I EXERCISES

1. _____   2. _____

Remember that the size of the angles of a regular polygon having n sides can be found by subtracting the mirror angle of the polygon, $\dfrac{360}{n}$, from 180°.

3.

Number of sides, n	3	4	5	6	7	9	10	12
Mirror angle	__	__	__	60	__	__	__	__
Angles of polygon	__	__	__	120	__	__	__	__

4. _____   5. _____

6. _____   7. _____

8. _____

_____

9. _____

The figures you will need for the rest of the exercises in this Set are provided in this workbook in Appendix B.

10. _____   11. _____

12. _____   13. _____

14. _____

15. _____

16. _____

17. _____

18. _____

19. _____

SET II EXERCISES

1. _____   2. _____   3. _____

4. _____   5. Yes or no?

6. Yes or no?   7. _____

8. _____

9. _____   10. _____

11. _____

12. _____

13. _____   14. _____

15. _____   16. _____   _____

SET III EXERCISES

1. Use tracing paper.

2. _____

3. _____

Reinforcing Past Lessons

Name the polygons with the following number of sides:

1. 6 _____

2. 5 _____

3. 4 _____

4. 10 _____

5. 7 _____

6. 12 _____

CHAPTER **5**
LESSON 4

Name _____

Date _____

SET I EXERCISES

1. _____   2. _____

3. _____   4. _____

5. _____

The figures you will need for the rest of the exercises in this Set are provided in this workbook in Appendix B.

6. _____   7. _____   8. _____   9. _____

10. _____   11. _____   12. _____   13. _____

14. _____   15. _____   16. _____   17. _____

18. _____   19. _____   20. _____   21. _____

22. _____   23. _____   24. _____   25. _____

26. _____   27. _____   28. _____   29. _____

30. _____   31. _____   32. _____   33. _____

34. _____

SET II EXERCISES

1. _____   2. _____   3. _____

4. _____   5. _____   6. _____

7. Exercise   1   2   3   4   5   6

   Number of polygons   ___ ___ ___ ___ ___ ___

8. _____   9. _____   10. _____

11. _____   12. _____

13. _____

14. _____

15.   Corners: _____

Edges: _____

16.   Corners: _____

Edges: _____

17.   Corners: _____

Edges: _____

18.   Corners: _____

Edges: _____

19.

Regular polyhedron	Number of faces	Number of corners	Number of edges
Tetrahedron	4	4	6
Cube	___	___	___
Octahedron	___	___	___
Dodecahedron	___	___	___
Icosahedron	___	___	___

20. _____

21. _____

22. _____

23. _____

Using the Calculator

Use your graphing calculator to graph the following function. Use the trace function to complete the table. Round each *y*-value to the nearest tenth. Use the standard screen of *x* {–10, 10}, *y* {–10, 10}. After graphing on your calculator, copy onto the graph paper provided on page 102. Be sure to record the scales for *x* and *y*. (In other words, are you counting by ones, twos, etc., on your graph?)

$$y = \frac{3}{x} + 4$$

x	–8	–5	–1	–.6	–.4	0	.4	.6	1	5	8
y	___	___	___	___	–3.5	___	___	___	___	___	___

Or you may instead use a regular calculator and solve each *y* value and graph.

SET III EXERCISES

Drawing:

Reinforcing Past Lessons

1. What is the common ratio of the sequence below? _____
   3, 24, 192, 1536, . . .

2. What mathematical expression would you use to determine the 8th term of the sequence? _____   What is the 8th term? _____

3. Identify the name of each special sequence below and determine the missing terms.

   a.   1   1   2   3   5   8   13   21   ____   ____   ____
      Name of sequence: _____

   b.   1   4   9   16   25   ____   ____   ____
      Name of sequence: _____

   c.   1   8   27   64   ____   ____   ____
      Name of sequence: _____

4. The logarithm of 10^5 is _____

5. The binary numeral for 107 is _____

6. The 7th term of the sequence of 4th powers in decimal form is _____

7. Write out the following term in words: 3×10^{27} _____

8. What is the base 10 number whose decimal logarithm is 6?
   Exponential form _____   Decimal form _____

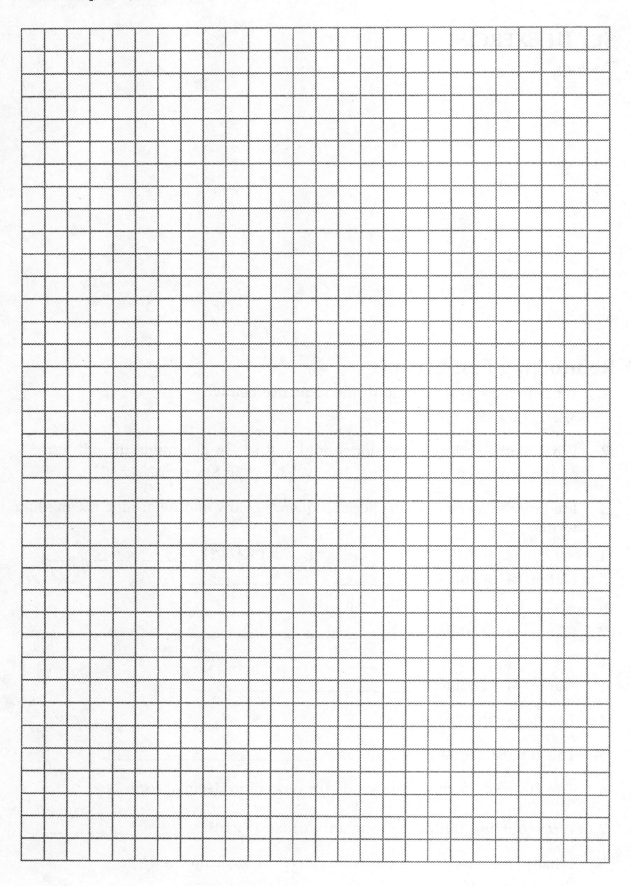

SET I EXERCISES

1. _____ 2. _____ 3. _____

4. _____ 5. _____ 6. _____

7. _____ 8. _____ 9. _____

10. _____ 11. _____ 12. _____

13. _____ 14. _____ 15. _____

16. _____ 17. _____ 18. _____

19. _____

20. _____

21. _____ 22. _____ 23. _____

24. _____ 25. _____ 26. _____

27. _____ 28. _____ _____

29. _____ 30. _____

31. _____ _____

32. _____ 33. _____

Using the Calculator

1. Use your graphing calculator to graph the following. Use the range of the screen for x {–5, 5} and y {–10, 10}. Transfer your graph to the graph paper provided on page 106 after tracing points and completing the table below.

$$y = x^3 - 6x$$

x	–3	–2	–1	0	1	2	3
y	__	__	__	__	__	__	__

You may instead use a regular calculator and solve each y value and graph. After graphing, answer the following questions.

a. Does the graph have line symmetry? Explain your answer.

_____

b. Does the curve have rotational symmetry? If yes, in what order?

_____

SET II EXERCISES

1. _____   2. _____

3. _____

4. _____

5. _____

6. Corners: _____

   Edges: _____

7. Corners: _____

   Edges: _____

8. Corners: _____

   Edges: _____

9. Corners: _____

   Edges: _____

10. _____

11. _____

12. _____

Reinforcing Past Lessons

1. The "mirror angle" of a regular polygon of n sides (in geometry this is called the "central angle") can be found by what formula?

   $m = $ _____

2. Using the formula above, determine the "mirror angle" of regular polygons with the given number of sides:

Number of sides, n	8	10	12	18	20
Angles of mirror, m	____	____	____	____	____

3. A pine tree's circumference decreased at a constant rate as a function of its height. Complete the table for the tree's circumference and its height. At the top of the tree, its circumference is zero. Determine the height of the tree.

    _____

c = circumference of the tree in feet, h = height of the tree in feet

c	42	39	36	33	30	__	__	__	__	__	__	__	__	__	__	__	__	__
h	1	2	3	4	__	__	__	__	__	__	__	__	__	__	__	__	__	__

4. A regular polyhedron is a solid having all faces in the shape of a

    _____

5. A solid with four equilateral triangles as faces is called a

    _____

6. A solid with eight equilateral triangles as faces is called a

    _____

7. A solid with 20 equilateral triangles as faces is called a

    _____

8. For base 2 logarithms, if the logarithm is 5, the number is _____.

9. For base 10 logarithms, if the logarithm is 8, the number is _____.

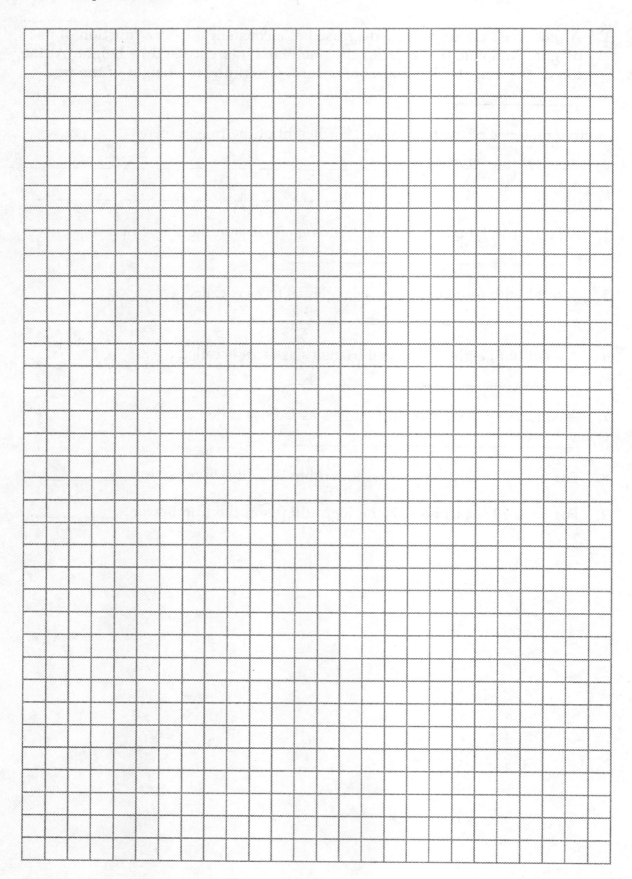

Name _____

Date _____

SET I EXERCISES

1 _____   2. _____

3. _____   4. _____

5. _____   6. _____

7. _____   8. _____

9. _____   10. _____

The "cut-outs" needed for the the following exercises are provided in Appendix B.

11. _____   12. _____

13. _____ 14. _____ 15. _____

16. _____

17. _____   18. _____

19. _____   20. _____

21. _____   22. _____

23. 2nd _____ 3rd _____ 4th _____

24. _____   25. _____

26. _____   27. _____

28. _____

SET II EXERCISES

1. _____ 2. _____ 3. _____

4. _____

5.

Pyramid	Number of sides in base	Number of faces	Number of corners	Number of edges
Triangular	____	____	____	____
Square	____	____	____	____
Pentagonal	____	____	____	____
Hexagonal	____	____	____	____

6. _____

7. _____   8. _____

9.

Prism	Number of sides in base	Number of faces	Number of corners	Number of edges
Triangular	____	____	____	____
Square	____	____	____	____
Pentagonal	____	____	____	____
Hexagonal	____	____	____	____

10. _____

11. _____   12. _____   13. _____

14. _____   15. _____   16. _____

17. _____

18. _____   19. _____   20. _____

21. _____

SET III EXERCISES

1. _____   2. _____

3. _____   4. _____

5. _____   6. _____

C H A P T E R 5
Summary and Review

Name _____

Date _____

SET I EXERCISES

1. _____

2. _____   3. _____

4. _____   5. _____

6. _____ 7. _____ 8. _____

9. _____ 10. _____

11. _____

12. _____   13. _____

14. _____   15. _____

16. _____   17. _____

18. _____

SET II EXERCISES

A B C D E F G H I K L M N O P Q R S T V W X Y Z

1. _____

2. _____   3. _____

4. _____   _____

5. Faces: _____   Corners: _____   Edges: _____

6. _____   _____

7.

8. _____   _____

9. _____   _____

10. _____   _____

11. _____   _____

12. _____   _____

13. _____   14. _____

15. _____   16. _____

17. _____   18. _____

19. _____

20. _____   21. _____

22. _____   _____

23. _____

24. _____

25. _____

SET III EXERCISES

1. _____

2. _____

3. _____

4. _____

Using the Calculator

Use your graphing calculator to draw designs that have symmetries. For each of the following graphs, be sure your calculator is in "parametric" mode. Use the standard range T, and Y, but change the x-min to -15, and x-max to 15. After determining rotational and line symmetries, make appropriate drawings on the axes provided on page 112, and draw the lines of symmetry.

1. $x_1 = 10 \cos 8T$ $y_1 = 10 \sin 8T$

a. Is there line symmetry? _____

b. Is there rotational symmetry? If yes, what is the order?

_____

2. $x_1 = 10 \cos 16T$ $y_1 = 10 \sin 16T$

a. Is there line symmetry? _____

b. Is there rotational symmetry? If yes, what is the order?

_____

3. $x_1 = 10 \cos 21T$ $y_1 = 10 \sin 21T$

a. Is there line symmetry? _____

b. Is there rotational symmetry? If yes, what is the order?

_____

4. $x_1 = 10 \cos 24T$ $y_1 = 10 \sin 18T$

a. Is there line symmetry? _____

b. Is there rotational symmetry? If yes, what is the order?

_____

1.

2.

3.

4.

Name _____

Date _____

SET I EXERCISES

Use the graph paper provided to complete Parts 1 and 2.

1. _____ 2. _____ 3. _____

4. _____ 5. _____

6.

Ellipse	Length of loop used	Distance between foci	Sum of distances of each point on ellipse from foci
A	32	12	20
B	___	___	___
C	___	___	___
D	___	___	___
E	___	___	___

7. _____

8. _____

9. _____

10. _____

Reinforcing Past Lessons

1. Complete the table for the function $y = 3x - 12$

x	−3	−2	−1	0	1	2	3	4
y	___	___	___	___	___	___	___	___

a. On the graph paper provided, draw an axis and graph the equation.

2. Complete the table and plot the curve on the graph for the function $y = 7 - x^2$.

x	−3	−2	−1	0	1	2	3
y	___	___	___	___	___	___	___

a. What is the name of the curve? _____

SET II EXERCISES

1. _____

2.

Ellipse	Distance from F_1	Distance from F_2
A	___	___
B	___	___
C	___	___
D	___	___
E	___	___

3. _____   4. _____   5. _____

6. _____

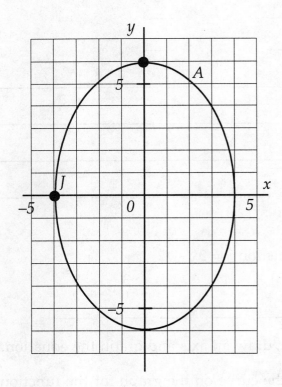

$$\frac{x^2}{16} + \frac{y^2}{36} = 1$$

7. _____

8. _____   9. _____

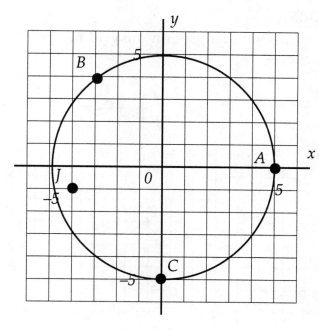

$$\frac{x^2}{25} + \frac{y^2}{25} = 1$$

10. _____

_____

11. _____

_____

12. _____

_____

13. _____   14. _____   15. _____

16. _____   17. _____

18. _____   _____

SET III EXERCISES

1. Stanton Drew: _____

Daviot: _____

2. _____ ; _____

3. _____

Supplemental Exercises

For the following equations graph both ellipses on the grid below. Use values of x and y to help you plot the curves.

1. $\dfrac{x^2}{4} + \dfrac{y^2}{36} = 1$

2. $\dfrac{x^2}{16} + \dfrac{y^2}{25} = 1$

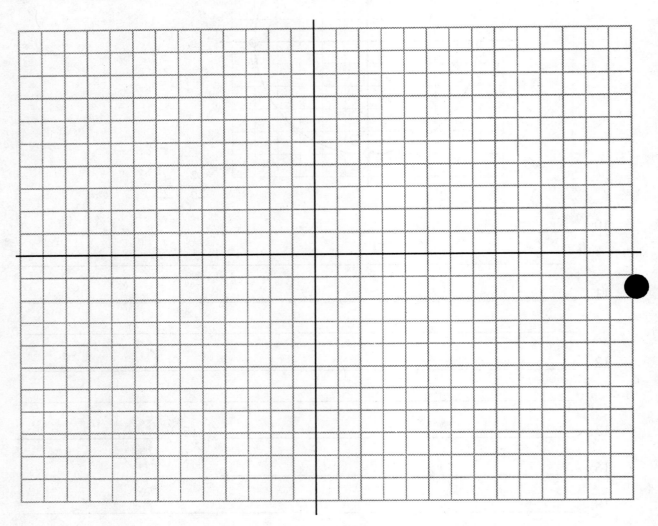

SET I EXERCISES

1. _____   2. _____

3. _____

4. _____

5. _____   6. _____

7.

x	0	1	2	4	5	6	7	8	9	10	11	12
y	0	11	20	—	—	—	—	—	—	—	—	—

8.

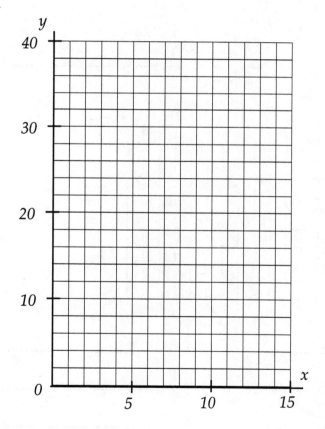

x	0.5	1.5	2.5	3.5	4.5	5.5	6.5	7.5	8.5	9.5	10.5	11.5
y	5.75	15.75	23.75	29.75	33.75	35.75	35.75	33.75	29.75	23.75	15.75	5.75

9.

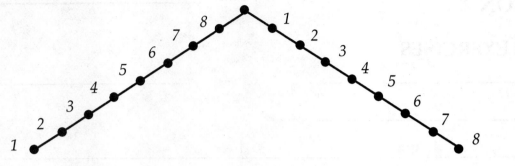

10. _____

11. _____

12. _____

SET II EXERCISES

1. Formula: $y = x^2$

x	–7	–6	–5	–4	–3	–2	–1	0	1	2	3	4	5	6	7
y	49	__	__	__	__	__	__	__	__	__	__	__	__	__	__

2.

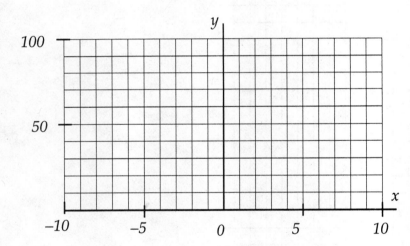

3. _____

4. _____

5. _____

6. _____

7. _____

8. _____

9. _____

SET III EXERCISES

Use the graph paper provided to complete Parts 1 and 2.

1. _____

2. _____

3. _____

SET I EXERCISES

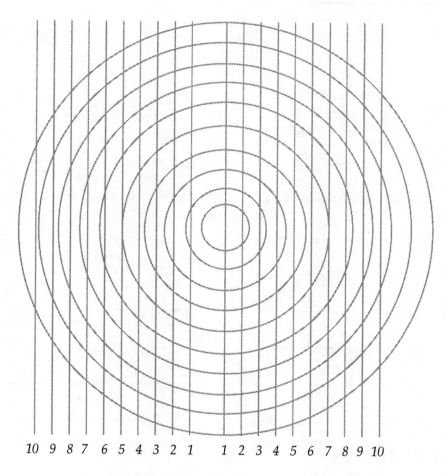

10 9 8 7 6 5 4 3 2 1　1 2 3 4 5 6 7 8 9 10

(For Part 1, rather than tracing the figure in the textbook, you can just mark the dots on the figure provided above.)

1. _____

2. _____

SET II EXERCISES

1. _____

2.

Point	Distance from F_1	Distance from F_2
A	___	___
B	___	___
C	___	___
D	___	___

3. _____   4. _____   5. _____

6. _____

7. _____

8. _____

9. _____

10. _____

11. _____

12. _____   13. _____

SET III EXERCISES

1. _____   2. _____

CHAPTER **6**

LESSON 4

Name _____

Date _____

SET I EXERCISES

1.

x	0°	15°	30°	45°	60°	75°	90°
sine of x	0	0.26	___	___	___	___	___

2. _____

3.

x	105°	120°	135°	150°	165°	180°
sine of x	___	___	___	___	___	___

4. _____

5.

x	195°	210°	225°	240°	255°	270°
sine of x	___	___	___	___	___	___

6. _____

7.

x	285°	300°	315°	330°	345°	360°
sine of x	___	___	___	___	___	___

8.

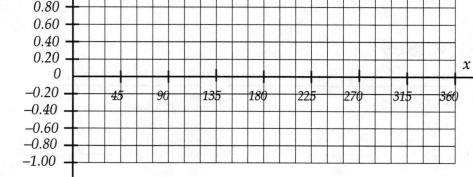

9. _____   10. _____

11. _____   12. _____

13. _____   14. _____

15. _____

Using the Calculator

1. Use a graphing calculator in "function" mode to graph the following equation. Use the range of x: {–1, 10} and y: {–10, 10}. Trace several points and transfer the graph to the grid provided. Label your points and scale.

$y = 6.8 \sin 1.3x$

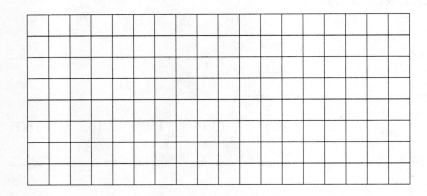

2. Write a sine curve equation of your own and graph it below using the graphing calculator.

$y = $ _____

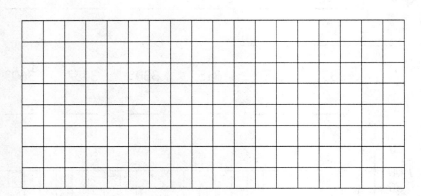

SET II EXERCISES

1.
Equation	Amplitude	Frequency	Wavelength
$y = 4 \sin 2x$	_____	_____	_____
$y = 2 \sin 5x$	_____	_____	_____
$y = 3 \sin \frac{1}{2}x$	_____	_____	_____

2. _____   3. _____   4. _____

5. _____   6. _____   7. _____

8. _____   9. _____   10. _____

11. _____   12. _____   13. _____

14. _____

SET III EXERCISES

1. _____   2. _____

3. _____   4. _____

5. _____

Supplemental Exercises

Draw the following sine curves on the axes provided.

1. $y = 4 \sin 3x$

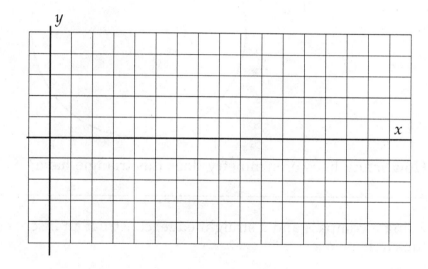

2. $y = 2 \sin 4x$

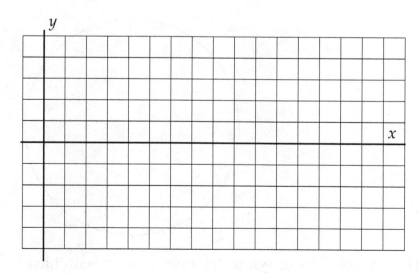

Reinforcing Past Lessons

These questions refer back to material you learned in Lessons 1 and 2 of Chapter 5.

1. Using a compass and a straight edge, construct an inscribed triangle within the circle below.

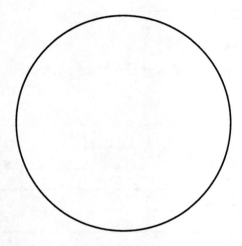

a. How many lines of symmetry does this triangle have?

_____

2. Using a compass and a straight edge, construct an inscribed hexagon with the circle below.

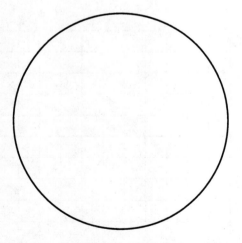

a. How many lines of symmetry does this hexagon have?

_____

Name _____

Date _____

SET I EXERCISES

1.

Angle	0°	30°	60°	90°	120°	150°	180°
Distance	0	0.55	1	___	___	___	___

2.

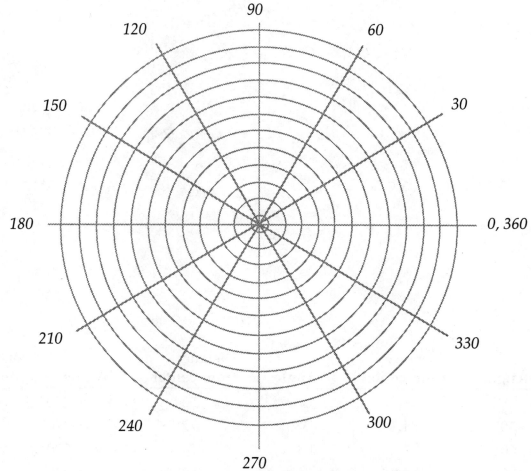

(For Exercise 2, rather than tracing the figure in the book, mark the dots directly onto the figure provided above.)

3. _____   4. _____

5.

Number of revolutions, x	0	1	2
Distance from center, y	0	___	___

6. _____   7. _____ , _____

8.

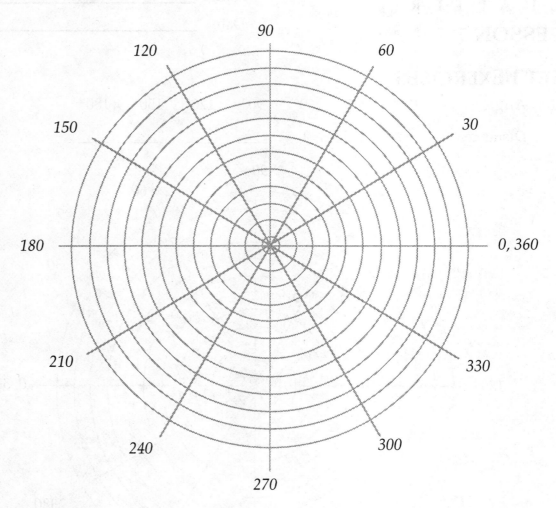

Angle	Distance from center	Angle	Distance from center	Angle	Distance from center
0°	1	300°	2.5	1 rev. + 240°	6.2
30°	1.1	330°	2.7	1 rev. + 270°	6.8
60°	1.2	1 rev.	3	1 rev. + 300°	7.5
90°	1.3	1 rev. + 30°	3.3	1 rev. + 330°	8.2
120°	1.4	1 rev. + 60°	3.6	2 revs.	9
150°	1.6	1 rev. + 90°	3.9	2 rev. + 30°	9.9
180°	1.7	1 rev. + 120°	4.3	2 rev. + 60°	10.8
210°	1.9	1 rev. + 150°	4.7	2 rev. + 90°	11.8
240°	2.1	1 rev. + 180°	5.2		
270°	2.3	1 rev. + 210°	5.7		

9. Number of revolutions, x 0   1   2

Distance from center, y 0   ____   ____

10. _____ 11. _____

SET II EXERCISES

1.   5   10   _____ _____ _____

2. _____   3. _____

4. _____

5.   2   4   _____ _____

6. _____   7. _____

8. _____

9.   3   6   _____ _____ _____ _____ _____ _____

10. _____   11. _____

12. _____

13.   . 4   _____ _____

14. _____   15. _____

16. _____   17. _____

18. _____

SET III EXERCISES

1. _____

2. _____

Supplemental Exercises

1. The coils of a burner on an electric stove form a spiral in which the distance between the coils is constant. Given the distance in centimeters from the center of the burner for part of the sequence, complete the sequence.

     1.8   3.0   4.2   _____ _____ _____

a. What type of sequence is this? _____

b. What type of spiral is this? _____

Using the Calculator

1. Use a graphing calculator to graph the following equations. Use the
    "parametric" mode. Trace several points and transfer your graph to the grid
    provided below. Enter the following into "$y =$":
    $x = T \cos T$ $y = T \sin T$

    Use the following range:
    T min = 0
    T max = 20
    T step = .105
    x min = −30
    x max = 30
    x scl = 1
    y min = −20
    y max = 20
    y scl = 1

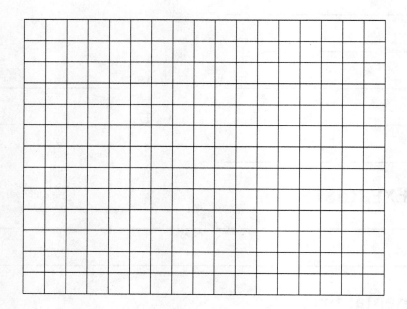

a. What type of curve is this? (Be specific.) _____

C H A P T E R 6
LESSON 6

Name _____

Date _____

SET I EXERCISES

1. _____

2. _____

3. _____

4. _____   5. _____

6. _____

7. _____

SET II EXERCISES

1. _____

2. _____

3. _____

4. _____

5. _____

6. _____

7. _____

8. _____

9. _____

10. _____   11. _____

12. _____

13. _____

14. _____   15. _____

16. _____

17. _____

18. _____

SET III EXERCISES

_____

_____

Using the Calculator

1. Use a graphing calculator to graph the following equations. Use the "parametric" mode. Transfer your graph to the grid provided below by tracing and labeling several points. Enter the following into "y =":

    $x1 = 3 (\cos T)^3$ $y1 = 3 (\sin T)^3$

    Use the following range:

    T min = 0
    T max = 6.28
    T step = .105
    x min = −4
    x max = 4
    x scl = .1
    y min = −4
    y max = 4
    y scl = 1

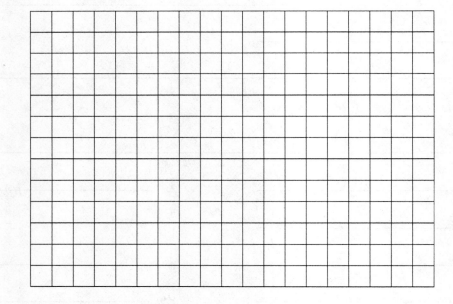

    a. What curve does this appear to be? _____

C H A P T E R 6
Summary and Review

Name _____

Date _____

SET I EXERCISES

1. 1 _____   2 _____

    3 _____   4 _____

2. _____   3. _____

4. _____   5. _____

6. _____   7. _____

8. _____

9. _____ 10. _____ 11. _____ 12. _____

13. _____   14. _____

15. _____   16. _____

SET II EXERCISES

1. _____

2. _____

3. _____   4. _____

5. _____   6. _____

7. _____

8. _____

9. _____

10. _____

11. Use tracing paper.

12. _____

13. _____

14.

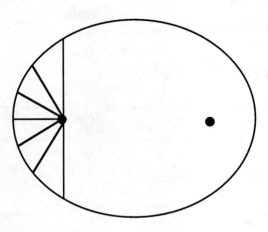

15. _____   16. _____

17. _____   18. _____

19. _____

SET III EXERCISES

1. _____   2. _____

3. _____   4. _____

5. _____   6. _____

Reinforcing Past Lessons

1. Sketch the following curve on the axis provided below.

$$\frac{x^2}{25} + \frac{y^2}{16} = 1$$

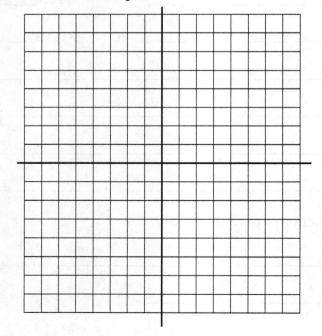

a. What curve is this? _____

SET I EXERCISES

1. _____ 2. _____ 3. _____

4. __ × __ × __ = __ 5. _____ 6. _____

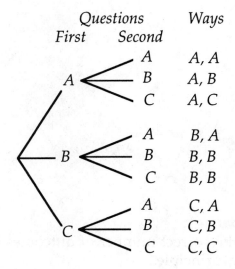

7. _____ 8. _____ 9. _____ = ___

10. _____ = ___ 11. _____ = ___

12. _____ = ___ 13. _____ = ___

14. _____ 15. _____ = ___

16. _____ = ___ 17. _____ = ___

18. _____ = ___ 19. _____ = ___

20. _____ = ___ 21. _____ = ___

22. _____ = ___ 23. _____ = ___

24. _____ = ___

Supplemental Exercises

1. Draw a tree diagram that shows all the possible answers for the first three questions of a true/false test. The diagram has been started for you. List the outcomes in the right-hand column.

First *Question*	*Second* *Question*	*Third* *Question*	*Outcomes*

True

False

a. Verify that you have the correct number of outcomes by using the fundamental counting principle.

_____ = _____

2. The state of Indiana makes license plates with the format: two digits, a letter, then four digits. (Remember that there are 10 digits in total, 0 through 9.) Use the fundamental counting principle to determine the number of license plates of this format that can be produced in the state of Indiana.

_____ = _____

3. The format of most Michigan license plates is two letters and four digits. Use the fundamental counting principle to determine how many of these plates could be produced.

_____ = _____

4. Consider another type of license plate that has seven places. The first two places are for digits, but no digits lower than 3 can be used in these places. The next place is a letter, but it must be before "h" in the alphabet. The last four places can be any digits. Use the fundamental counting principle to determine how many of these types of plates could be produced.

_____ = _____

SET II EXERCISES

Symbol	Reel 1	Reel 2	Reel 3
Bar	3	2	1
Bell	1	5	8
Cherry	2	6	0
Lemon	0	0	4
Melon	2	2	2
Orange	5	5	4
Plum	7	3	3
7	1	1	1
(Positions)	20	20	20

1. _____   2. _____

3. _____ = _____   4. _____ = ___

5. _____ = _____   6. _____

7. _____ = _____

8. _____ = _____   9. _____ = _____

10. _____ = _____   11. _____ = _____

12. _____ = _____   13. _____ = ___

14. _____ = _____

15. _____ = _____   16. _____

17. _____   18. _____

19. _____ = _____

20. _____ = _____

21. _____ = _____   22. _____

23. _____ = _____   24. _____ = ___

25. _____ = _____

26. _____ = _____

27. _____ = _____   28. _____

29. _____

_____

SET III EXERCISES

_____

_____

Supplemental Exercises

There are four roads that lead from Southtown to Uptown, and three roads
that lead from Uptown to Northtown. In the space below, make a tree diagram
of the road possibilities. Use the fundamental counting principle to determine
the number of ways in which one could make the trip from Southtown to
Northtown.

_____ = _____

SET I EXERCISES

A *permutation* is defined as an arrangement of things in a definite order. The permutation of eight things taken three at a time is written $_8P_3$ where "P" stands for permutations.

The permutation of eight things taken eight at a time is written $_8P_8$. It can also be written 8!. The exclamation point is called the *factorial symbol* and, in this case, means to multiply the consecutive counting numbers from 8 down to 1, or

$$8 \times 7 \times 6 \times 5 \times 4 \times 3 \times 2 \times 1.$$

1. _____   2. _____

3. _____

A E I P R S T

4. _____   5. _____

6. _____   7. _____

8. _____

9. _____   10. _____

11. _____   12. _____

13. _____   14. _____

15. _____

16. _____

17. _____   18. _____

19. _____   20. _____

21. _____

Reinforcing Past Lessons

1. Make a tree diagram for the toss of two coins in the space below.

First
Coin

Second
Coin

Outcomes

a. Verify that you have the correct number of outcomes by using the fundamental counting principle.

_____ = _____

SET II EXERCISES

1! = 1	9! = 362,880
2! = 2	10! = 3,628,800
3! = 6	11! = 39,916,800
4! = 24	12! = 479,001,600
5! = 120	13! = 6,227,020,800
6! = 720	14! = 87,178,291,200
7! = 5,040	15! = 1,307,674,368,000
8! = 40,320	

1. _____ because _____

2. _____ because _____

3. _____ because _____

4. _____ because _____

5. _____ because _____

6. $_8P_3$ = _____

7. $_{11}P_1$ = _____

8. $_9P_5$ = _____

9. $\dfrac{8!}{5!} =$ _____

10. $\dfrac{11!}{10!} =$ _____

11. $\dfrac{9!}{4!} =$ _____

12. _____   13. _____

14. _____   15. _____

16. _____

> 1. However, nobody had seen one for months.
> 2. He thought he saw a shape in the bushes.
> 3. Mark had told him about the foxes.
> 4. John looked out the window.
> 5. Could it be a fox?

17. _____   18. _____

19. _____

20. _____

SET III EXERCISES

> 7:00   "Studio One"
> 8:00   "Burns and Allen"
> 8:30   "Talent Scouts"
> 9:00   "I Love Lucy"
> 9:30   "December Bride"
> 10:00   "Mr. District Attorney"
> 10:30   News

1. _____

2. _____

Supplemental Exercises

1. In the manner of the Exercises 1–5 in Set II, *show whether* each of the following equations is true or false.

a. $10! - 1! = 9!$

_____ because _____

b. $5! \times 3! = 6!$

_____ because _____

c. $5! \times 6 \times 7 = 7!$

_____ because _____

2. In how many ways can 6 people line up for a theater box office?

_____

3. In how many ways can the letters in the word "toes" be scrambled?

_____

4. If there are 10 horses running in a race, in how many ways can first and second place be taken?

_____

5. In how many ways can the letters in the word "Vancouver" be arranged?

_____

6. In how many different ways can 3 men and 3 women be seated at a table if men and women must be seated alternately (man/woman/man/woman etc.)? Try using the fundamental counting principle and then multiplying your choices.

_____

● C H A P T E R **7**
LESSON 3

Name _____

Date _____

SET I EXERCISES

1. _____   2. _____

3. _____   4. _____

5. _____

6. _____

7. _____

8. _____

9. _____

10. _____   11. _____

12. _____

● 13. _____

14. _____   15. _____

16.

         |ıılı

17. _____ ; _____

18. _____ ; _____

SET II EXERCISES

1. _____   2. _____

3. _____   4. _____

5. _____   6. _____

● 7. _____   8. _____

9. _____   10. _____

11.

Number of heads	10	9	8	7	6	5	4	3	2	1	0
Number of tails	0	1	2	3	4	5	6	7	8	9	10
Number of orders	1	10	__	__	__	__	__	__	__	__	__

12. _____   13. _____

14. _____   15. _____

16. _____

17. _____

SET III EXERCISES

_____

_____

Supplemental Exercises

1. In how many different ways can the letters "deeds" be arranged? Show the mathematical expression with factorials and find the value of the expression.

    _____

2. In how many ways can the letters in the word "Alaska" be arranged?

    _____

Reinforcing Past Lessons

1. A guardrail along a highway is in the shape of a certain curve you studied in Chapter 6. The equation is given for the curve. Draw it on the axis provided, for two-and-a-half wavelengths.

   Equation: $y = 2 \sin x$

Name _____

Date _____

SET I EXERCISES

1. _____   2. _____

3. _____   4. _____

5. _____

6. _____   7. _____

8. _____   9. _____

10. _____   11. _____

12. _____   13. _____

14. _____   15. _____

16. _____

17. Number of raised dots   1   2   3   4   5   6
     Number of ways   ___   ___   ___   ___   ___   ___

18. _____   19. _____

20. _____   21. _____

22. _____   23. _____

24. _____   25. _____

SET II EXERCISES

1. _____

2. _____

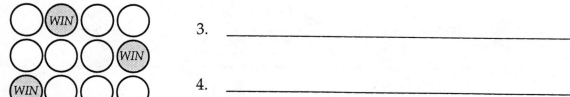

3. _____

4. _____

5. _____   6. _____

7. _____   8. _____

9. _____

10. _____

11. _____

12. _____

13. Draw your solution below.

14. _____

SET III EXERCISES

1. _____

2. _____

3. _____

4. _____

5. _____

6. _____

Supplemental Exercises

1. Eight people are waiting on standby for a flight. Four seats become available: one with ample leg room, one window seat, one aisle seat, and one seat in the middle. The people on standby will be given their choice of seats in the order in which their names are picked from the list.

   a. How many arrangements of people on standby can take the available seats on the flight?

   _____

   b. Is this a combination or a permutation?   _____

   c. Why?   _____

2. One part of a mathematics test contains eleven problems. Students are required to work any eight problems of their choice. How many selections of eight problems are possible?

   _____

3. How many two-digit numbers can be written using only the digits 5 and 6, where repetition is allowed?

   _____

   a. Prove your answer by listing all the possibilities.

4. How many five-digit numbers can be formed from the digits 3, 4, 5, 6, and 7 if repetition of digits is allowed?

   _____

5. A car dealership offers its latest model with a choice of five exterior colors, three interior colors, and the option of a sun roof. How many different cars does a buyer have to choose from?

   _____

   a. What method of probability did you use to determine your answer?

   _____

6. For its first time subscribers, a book club offers 3 books for $1 with a choice of 22 book titles. How many different sets of 3 books can a prospective member choose?

   _____

a. Is the order of selection important for this problem? Why or why not?

    _____ ; _____

    _____

b. Is this a combination or a permutation? _____

7. Complete the following table showing the ways in which a true/false test having different numbers of questions can be answered.

Number of questions	1	2	3	4
Number of ways for answers to appear	__	__	__	__

a. Are the choices made in answering the questions of a true/false test dependent or independent? _____

b. By which method did you determine your answers in this problem: the fundamental counting principle, permutations, or combinations?

    _____

SET I EXERCISES

1. _____   2. _____

Nation	First stripe	Second stripe	Third stripe
Belgium	black	yellow	red
Chad	blue	yellow	red
France	blue	white	red
Guinea	red	yellow	green
Ireland	green	white	gold
Italy	green	white	red
Ivory Coast	gold	white	green
Mail	green	yellow	red
Nigeria	green	white	green
Peru	red	white	red
Romania	blue	gold	red

Black Blue Gold Green Red	Gold White Yellow	Gold Green Red

3. _____

Black Blue Gold Green Red White Yellow	Black Blue Gold Green Red White Yellow	Black Blue Gold Green Red White Yellow

4. _____

5. _____ 6. _____

7. _____ 8. _____

9. _____ 10. _____

11. _____ 12. _____

13. _____ 14. _____

15. _____ 16. _____

SET II EXERCISES

1. _____ 2. _____

3. _____ 4. _____

5. _____ 6. _____

7. _____ 8. _____

9. _____ 10. _____

11. _____ 12. _____

13. _____ 14. _____

SET III EXERCISES

1. _____

2. _____

SET I EXERCISES

When working problems with probability, your final fractional answer should be in simplest form. For example $\frac{2}{4}$ is equal to $\frac{1}{2}$ in simplest form.

1. _____ 2. _____ 3. _____ 4. _____

5. _____ 6. _____ 7. _____ 8. _____

9. _____ 10. _____ 11. _____ 12. _____

13. _____ 14. _____ 15. _____ 16. _____

17. _____ 18. _____ 19. _____

20. _____

21. _____ 22. _____ 23. _____ 24. _____

25. _____ 26. _____

27. _____

28. _____ 29. _____ 30. _____

SET II EXERCISES

1. _____ 2. _____

3. _____ 4. _____

5. _____ 6. _____ 7. _____ 8. _____

9. _____

10. _____ 11. _____

12. _____ 13. _____

14. _____ 15. _____

16. _____ ; _____

SET III EXERCISES

Spot aimed at	Resulting Score									
	60	30	20	15	10	5	3	2	1	0
A	33	0	87	24	0	69	21	0	66	0
B	12	18	144	6	3	57	6	3	51	0
C	0	57	123	0	15	33	0	12	30	30

1. _____ 2. _____ 3. _____ 4. _____

5. _____ 6. _____ 7. _____ 8. _____

9. _____ 10. _____ 11. _____ 12. _____

13. _____

14. _____

Supplemental Exercises

1. A man has 11 pairs of socks in his sock drawer: 3 brown, 6 black, and 2 blue.

   a. If the man reaches into his drawer with his eyes closed, what is the fractional probability that the pair of socks he chooses is black?

   _____

   b. What is the probability that the pair he chooses is brown?

   _____

   c. What is the probability that the pair he chooses is blue?

   _____

2. There are five possible answers to a multiple-choice question, and you really don't have the slightest idea which one is correct. If you just pick an answer at random, what is the probability that you will choose the correct answer?

   Fraction: _____ Percent: _____

Name _____

Date _____

SET I EXERCISES

1. | Sum of two dice | 2 | 3 | 4 | 5 | 6 | 7 | 8 | 9 | 10 | 11 | 12 |
    |---|---|---|---|---|---|---|---|---|---|---|---|
    | Number of ways | 1 | __ | __ | __ | __ | 6 | __ | __ | __ | 2 | __ |

2. _____

3. _____ , _____

4. _____

5. _____ , _____

6. _____

7. _____ , _____

8. _____

9. _____

10. _____

11. _____

12. _____

13.

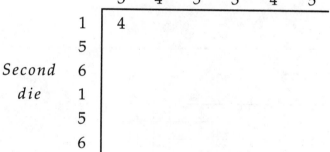

First die

		3	4	5	3	4	5
	1	4					
	5						
Second	6						
die	1						
	5						
	6						

14. _____

15. _____

16. _____

17. _____

18. _____

Supplemental Exercises

1. What is the probability on the roll of two dice that the sum will be divisible by 5?

   Fraction: _____   Percent: _____

2. What is the probability on the roll of two dice that the sum will be divisible by 4?

   Fraction: _____   Percent: _____

SET II EXERCISES

1. _____

2. _____   3. _____

4. Sum of three dice 3 4 5 6 7 8 9 10 11 12 13 14 15 16 17 18
    Number of ways   _ _ 6 _ _ _ _ _   _   _   _   _   _   _   _   _

5. _____   6. _____

7. _____

8. _____   9. _____   10. _____   list: _____

11. _____   12. _____   13. _____

14. _____

15. _____   16. _____   17. _____   18. _____

19. _____   20. _____

21. _____

22.

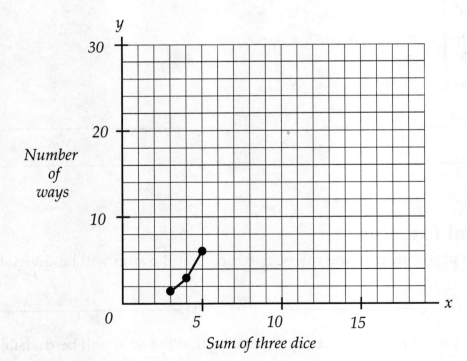

23. _____   24. _____   25. _____

26. _____

SET III EXERCISES

1. _____

2.

		Die B					
		0	1	7	8	8	9
Die C	5						
	5						
	6						
	6						
	7						
	7						

3. _____

4.

		Die C					
		5	5	6	6	7	7
Die D	3						
	4						
	5						
	5						
	11						
	12						

5. _____

6.

		Die D					
		3	4	4	5	11	12
Die A	1						
	2						
	3						
	9						
	10						
	11						

7. _____

8. _____
_____

Reinforcing Past Lessons

1. Think back to what you learned in Chapter 3, Lesson 6 for this exercise. "Wind chill" is a measure of the effect of temperature and wind on the human body. The combination of wind and the cold temperature make the body feel colder than the effect of the cold temperature alone. Make a graph of the wind chill effect on a 20 mile per hour wind by plotting the values given and connecting those points with a smooth curve. (Temperatures are in degrees Fahrenheit.)

Actual temperature	30	20	10	0	–10	–20
Wind chill (with 20 mph wind)	4	–10	–24	–39	–53	–67

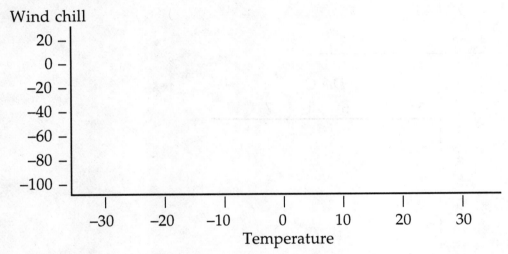

What would the wind chill be at these temperatures?

a. –30 _____

b. 5 _____

c. 25 _____

d. For which of these temperature(s) did you use interpolation? _____

e. For which of these temperature(s) did you use extrapolation? _____

2. A restaurant offers a choice of 5 appetizers, 6 main courses, and 4 desserts. How many different meals coud be ordered in this restaurant?

Name _____

Date _____

SET I EXERCISES

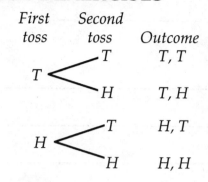

First toss | Second toss | Outcome
T → T : T, T
T → H : T, H
H → T : H, T
H → H : H, H

1. _____   2. _____

3. _____   4. _____

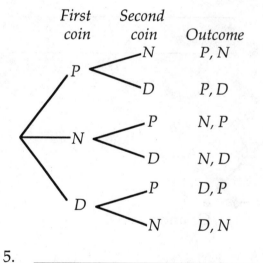

First coin | Second coin | Outcome
P → N : P, N
P → D : P, D
N → P : N, P
N → D : N, D
D → P : D, P
D → N : D, N

5. _____   6. _____

7. _____   8. _____

9. _____   10. _____

11. _____   12. _____

13. _____   14. _____

15. _____   16. _____

17. _____   18. _____

19. _____   20. _____

21. _____   22. _____

23. _____   24. _____

SET II EXERCISES

1. _____   2. _____

3. _____   4. _____

5. _____   6. _____

7. _____   8. _____

9. _____   10. _____

11. _____   12. _____

13. _____   14. _____

Anaheim
Atlanta
60%
70%
80%
Dallas
75%
Orlando

15. _____   16. _____

17. _____ 18. _____

19. _____   20. _____

SET III EXERCISES

1. _____   2. _____

3. _____   4. _____

5. _____   6. _____

7. _____

Using the Calculator

Use the graphing calculator to graph $y = -4x^2 + 15.3$.
Use the range of $x\{-10, 10\}$ and $y\{-10, 20\}$. Trace at least four points and transfer your curve to the graph provided below. Be sure to indicate the scale being used. Label the equation.

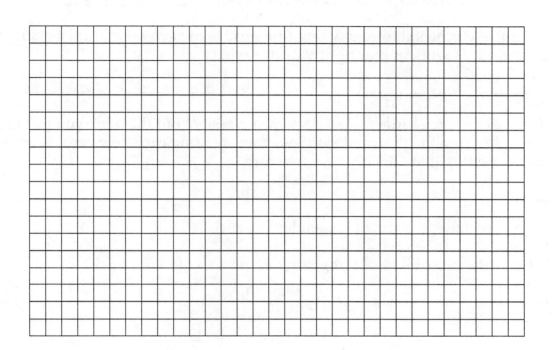

Supplemental Exercises

For each of the following problems show a) the fractional probability, and b) the percent probability.

1.   What is the probability of rolling a die twice and rolling a five both times?

a.   _____   b.   _____

2.   What is the probability of drawing two cards from a standard deck in succession without putting any cards back, and obtaining two queens?

a.   _____   b.   _____

3.   What is the probability of drawing three black cards from a standard deck in succession without putting any cards back?

a.   _____   b.   _____

c.   Is this an independent or dependent event?   _____

4. If you draw three cards from a standard deck in succession, without putting the cards back into the deck after each selection, what is the probability that each card is a face card?

a. _____   b. _____

c. Is this an independent or dependent event? _____

5. What is the probability of drawing four spades in succession from a standard deck of cards without putting any cards back?

a. _____   b. _____

6. If you draw three cards from a deck in succession, but this time you put the cards back into the deck and shuffle before choosing each new card, what is the probability that each card is a face card?

a. _____   b. _____

c. Is this an independent or dependent event? _____

7. What is the probability of **not** getting a 2 on a single roll of a die?

a. _____   b. _____

Name _____

Date _____

SET I EXERCISES

1. _____   2. _____

2.
Number of boys	0	1	2
Number of possibilities	1	—	—
Percent probability	25%	—	—

4. _____   5. _____

	First time at bat	Second time at bat	Third time at bat	Hits
		H	H	H, H, H
	H		O	H, H, O
		O	H	H, O, H
			O	H, O, O
		H	H	O, H, H
	O		O	O, H, O
		O	H	O, O, H
			O	O, O, O

6. _____   7. _____

8. _____   9. _____

10. _____   11. _____

12.
Number of hits	0	1	2	3
Number of possibilities	1	—	—	—
Percent probability	—	—	—	—

13. _____   14. _____

14. _____ 15. _____ 16. _____

17. _____

18. _____

19. _____

20. _____   21. _____

22. _____   23. _____

24. _____

25. _____   26. _____

27. _____

28. _____

29.
Number of days it snows	0	1	2	3	4
Number of possibilities	1	—	—	—	—
Percent probability	—	—	—	—	—

30. _____   31. _____   32. _____

33. _____

SET II EXERCISES

1. _____   2. _____

3. _____   4. _____

5. _____   6. _____

7. _____   8. _____

SET III EXERCISES

1. | Number of intersections at which path goes to the left | 0 | 1 | 2 | 3 | 4 | 5 |
    |---|---|---|---|---|---|---|
    | Number of possibilities | 1 | ___ | ___ | ___ | ___ | ___ |
    | Percent probability | ___ | ___ | ___ | ___ | ___ | ___ |

2. _____   3. _____

4. _____

5. _____

Supplemental Exercises

For Exercises 1, 2, and 3, show all your work, as well as your answers, which should be expressed as a percent figure. (Batting averages are in thousandths; a batting average of .325 means that a batter makes hits in 325 out of 1000 times at bat, or that the batter makes hits on 32.5 percent of his times at bat.)

1. Between 1907 and 1919, Ty Cobb of the Detroit Tigers had the highest batting average in the American League for every year but one. His best average during these years was .420, which he hit in 1911. In that year, what was the probability of Cobb getting three hits in three successive times at bat? *(See Exercises 6 and 7 in Set I for help, if necessary.)*

    _____

2. Stan Musial of the St. Louis Cardinals was the National League's batting champion six times between 1943 and 1957. His best batting average during these years was in 1948 when he hit .376. In that year, what was the probability of Musial getting four hits in four consecutive times at bat?

    _____

3. In the 1990–1991 basketball season, Larry Bird of the Boston Celtics had a three-point field goal percentage of .389, which translates to 38.9% success in his attempts to make three-point shots.

a. What is the probability that in four attempts at three-point shots, Bird would have made all four?

    _____

b. What is the probability that Bird would have made all five of three-point attempts?

    _____

4. Consider 4 tosses of a coin. What is the probability that at least three of the tosses turn up heads?

Using the Calculator

A hyperbola would result from the equation:

$$\frac{y^2}{200} - \frac{x^2}{150} = 1$$

In order to use the graphing calculator the equation(s) must be in the "$y =$ ___" form. Therefore, this equation has to be re-expressed as follows:

$$y_1 = \sqrt{\left(\frac{4}{3}\right)x^2 + 200}$$

$$y_2 = -\sqrt{\left(\frac{4}{3}\right)x^2 + 200}$$

Graph the equation on the graphing calculator. Use the range $x\{-20, 20\}$ and $y\{-30, 30\}$. Let x–scale = 5 and y–scale = 5. Trace at least 6 points and transfer to the graph provided below.

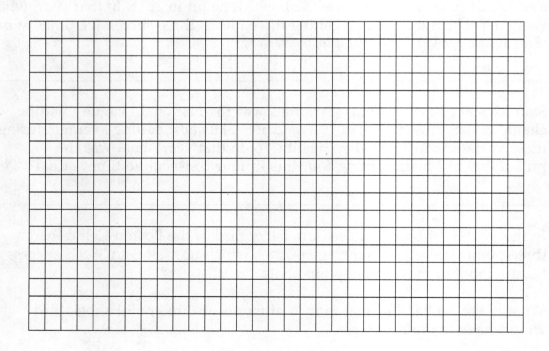

LESSON 5

SET I EXERCISES

Consider the probability that an event will happen or that it will not. The sum for these two types of events is 100%. For instance, if there is a 30% chance it will rain on a given day, there is a 70% chance it will not.

1.

Number of rows							Sum of numbers in row
1			1	1			2
2		1	2	1			4
3		1	3	3	1		8
4	1	4	6	4	1		16
5		___ ___ ___ ___ ___ ___					___
6		___ ___ ___ ___ ___ ___					___
7		___ ___ ___ ___ ___ ___					___
8		___ ___ ___ ___ ___ ___ ___					___
9		___ ___ ___ ___ ___ ___ ___					___
10		___ ___ ___ ___ ___ ___ ___ ___					___

2. _____

3.

Number of teenagers having an accident	1	2	3	4	5
Number of ways	___	___	___	___	___

4. _____   5. _____

6. _____   7. _____

8. _____   9. _____

10. _____   11. _____

12. _____

13. _____

14.

Number of baskets	0	1	2	3	4	5	6
Number of ways	___	___	___	___	___	___	___

15. _____

16. _____

17. _____   18. _____

19. _____   20. _____

21. Number of copies that work   0   1   2   3
    Number of ways   ___   ___   ___   ___

22. _____

23. _____

24. _____

25. _____

26. _____

27. Number of correct answers   0   1   2   3   4   5   6   7   8
    Number of ways   ___ ___ ___ ___ ___ ___ ___ ___ ___

28. _____   29. _____

30. _____   31. _____

32. _____

SET II EXERCISES

1. | Number of heads | 0 | 1 | 2 | 3 | 4 | 5 | 6 | 7 | 8 | 9 | 10 |
   |---|---|---|---|---|---|---|---|---|---|---|---|
   | Number of ways | 1 | 10 | ___ | ___ | ___ | ___ | ___ | ___ | ___ | ___ | ___ |
   | Percent probability | 0% | 1% | ___ | ___ | ___ | ___ | ___ | ___ | ___ | ___ | ___ |

2.

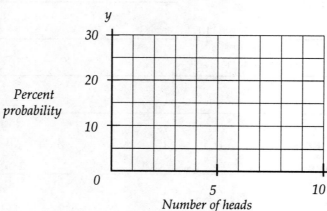

Number of heads

3. _____   4. _____

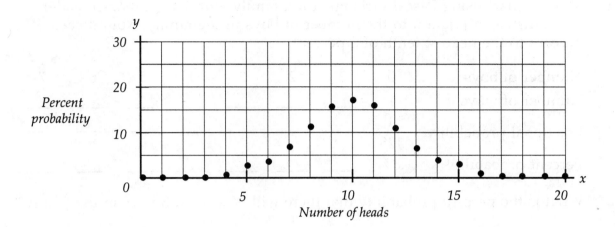

5. _____

6. _____

7. _____   8. _____

9. _____   10. _____

SET III EXERCISES

1. _____

2. _____

3. _____

4. _____

5. _____

6. _____

7. _____

8. _____

9. _____

Supplemental Exercises

1. Make a chart using Pascal's triangle for a family with 7 children. Consider the statistics in relation to the number of boys in the family. Round each percent to the nearest tenth of a percent.

Number of boys	0	1	2	3	4	5	6	7
Number of ways	1	___	___	___	___	___	___	___
Fractional probability	$\frac{1}{128}$	___	___	___	___	___	___	___
Percent probability	.8%	___	___	___	___	___	___	___

a. What is the percent probability that there will be at least 3 boys in the family?

b. What is the percent probability that there will be 7 boys in the family?

c. What is the fractional probability that there will be fewer than 4 boys in the family?

2. Three statisticians each take three coins and toss them into a circle, counting the total number of tails. After tossing all nine coins 100 times, they find they are very close to the theoretical outcome for heads and tails (the statistics that come from making a chart). Using Pascal's triangle, complete the chart for the tossing of 9 coins. Also determine the fractional and percent probability for each case.

Number of tails	0	1	2	3	4	5	6	7	8	9
Number of ways	1	___	___	___	___	___	___	___	___	___
Fractional probability	___	___	___	___	___	___	___	___	___	___
Percent probability	___	___	___	___	___	___	___	___	___	___

a. What is the fractional probability that at least 5 tails will turn up?

b. What is the percent probability that no tails will turn up?

c. What is the fractional probability that fewer than 4 tails will turn up?

SET I EXERCISES

The event that something happens and the event that it does not happen are called *complementary events*.

1. _____   2. _____

3. _____   4. _____

5. _____   6. _____

7. _____   8. _____

9. _____   10. _____

11. _____   12. _____

13. _____   14. _____

15. _____   16. _____

17. _____   18. _____

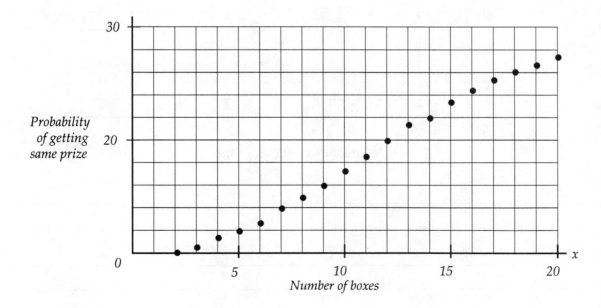

19. _____   20. _____   21. _____

SET II EXERCISES

1. _____   2. _____

3. _____

4. _____

5. _____

6. _____

7.
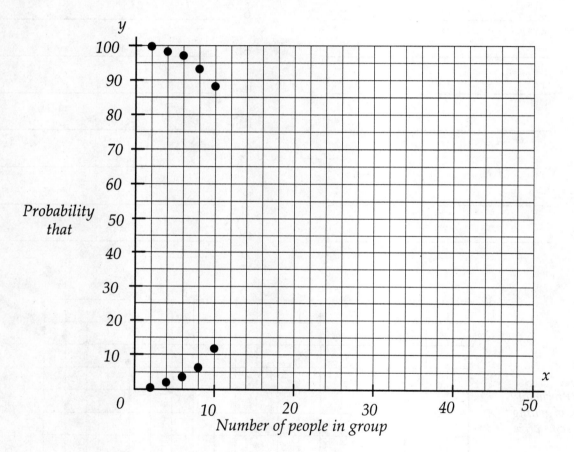

8. _____

   _____

9. _____

   _____

10. _____

11. _____

   _____

12. _____

13. _____

14. _____   15. _____

SET III EXERCISES

1. _____   2. _____

3. _____   4. _____

5. _____   6. _____

7. _____   8. _____

9. _____

Supplemental Exercises

1. The probabilities of two complementary events total what amount?

   _____

2. What is the probability that on a single toss of a die, you will get a 1?

   _____

3. What is the probability that on a single toss of a die, you will not get a 1?

   _____

4. What is the probability that on a single toss of a die, you will not get an odd
   number? _____

5. What is the probability that in choosing two cards from a deck at random
   (without putting the first card back), you will not get a king? _____

Reinforcing Past Lessons

1. List all the possible ways of using the letters in the word OHIO.

   a. Confirm your answer by showing how to determine the number of ways
      mathematically. _____

   b. Is this a permutation or a combination? _____

Using the Calculator

1. Remember what you learned in Chapter 3, Lesson 1 and complete this table for the formula $y = 12 - 2x$. Graph the table on the grid below. Check your answer with a graphing calculator.

x	−2	−1	0	1	2	3
y	__	__	__	__	__	__

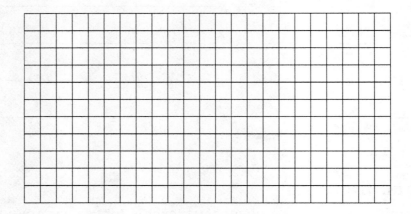

2. Now recall Chapter 6, Lesson 1, and use the graphing calculator to graph the following conic section. Trace and graph on the grid provided.

    Use the range of:
    $x\{-14, 14\}$
    $y\{-10, 10\}$

    $y = \sqrt{(81 - x^2)}$
    $y = -\sqrt{(81 - x^2)}$

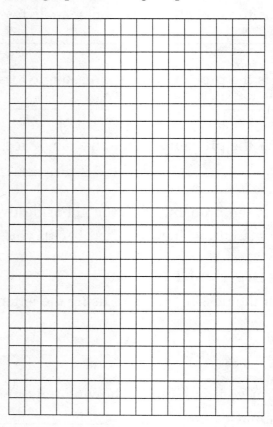

a. What is the name of this conic section? _____

SET I EXERCISES

1. _____

2. _____

A smoker wants to quit	66%
A smoker has tried to quit	84%
A smoker will succeed in quitting	21%

3. _____

4. _____

5. _____

6. _____

7. _____

8. _____

9. _____

10. _____

11. _____

12.

13. _____

14. _____

15. _____

16. _____

17. _____

18. _____

SET II EXERCISES

1. _____   2. _____   3. _____

4. _____   5. _____   6. _____

7. _____   8. _____

9. _____

10. _____

11. _____   12. _____

13. _____

14. _____   15. _____

16. _____

17. _____

18. _____

19. _____

20. _____

21. _____

SET III EXERCISES

1. _____   2. _____

3. _____   4. _____

5. _____

SET I EXERCISES

1. _____   2. _____   3. _____

4. _____   5. _____

6.

Distance in meters	Tally marks	Frequency
6.01–6.50	_____	____
6.51–7.00	_____	____
7.01–7.50	_____	____
7.51–8.00	_____	____
8.01–8.50	_____	____
8.51–9.00	_____	____

7.

Distance in meters	Tally marks	Frequency
6.01–6.50	_____	____
6.51–7.00	_____	____
7.01–7.50	_____	____
7.51–8.00	_____	____
8.01–8.50	_____	____
8.51–9.00	_____	____

8.

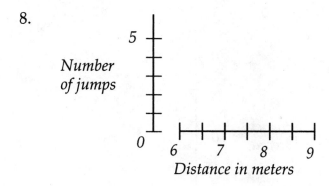

Number of jumps

Distance in meters

10. _____

11. _____   12. _____

13. _____   14. _____

15. _____

16. _____

SET II EXERCISES

1.

Age in years	Tally marks	Frequency
20–29	_____	____
30–39	_____	____
40–49	_____	____
50–59	_____	____
60–69	_____	____
70–79	_____	____
80–89	_____	____

2.

Age in years	Tally marks	Frequency
20–29	_____	____
30–39	_____	____
40–49	_____	____
50–59	_____	____
60–69	_____	____
70–79	_____	____
80–89	_____	____

3.

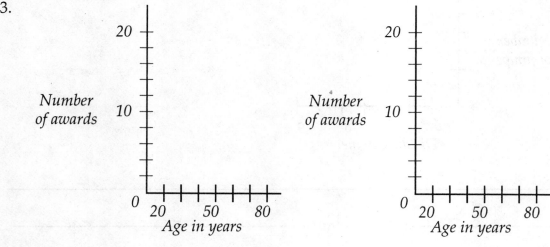

4.

Age in years	Tally marks	Frequency
20–29	_____	_____
30–39	_____	_____
40–49	_____	_____
50–59	_____	_____
60–69	_____	_____
70–79	_____	_____
80–89	_____	_____

5.

Age in years	Tally marks	Frequency
20–29	_____	_____
30–39	_____	_____
40–49	_____	_____
50–59	_____	_____
60–69	_____	_____
70–79	_____	_____
80–89	_____	_____

6.

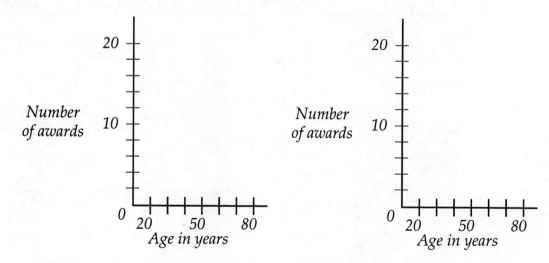

7. _____

_____

8. _____   9. _____

10. _____   11. _____

12.

Letter	Tally marks	Frequency	Letter	Tally marks	Frequency
A	_____	___	N	_____	___
B	_____	___	O	_____	___
C	_____	___	P	_____	___
D	_____	___	Q	_____	___
E	_____	___	R	_____	___
F	_____	___	S	_____	___
G	_____	___	T	_____	___
H	_____	___	U	_____	___
I	_____	___	V	_____	___
J	_____	___	W	_____	___
K	_____	___	X	_____	___
L	_____	___	Y	_____	___
M	_____	___	Z	_____	___

13. _____

_____

14.

Letter	Tally marks	Frequency	Letter	Tally marks	Frequency
___	_____	___	___	_____	___
___	_____	___	___	_____	___
___	_____	___	___	_____	___
___	_____	___	___	_____	___
___	_____	___	___	_____	___

15. _____

_____

16. _____ ; _____

17.

Letter	Tally marks	Frequency	Letter	Tally marks	Frequency
___	_____	___	___	_____	___
___	_____	___	___	_____	___
___	_____	___	___	_____	___
___	_____	___	___	_____	___
___	_____	___	___	_____	___

18. _____

_____

19. _____ ; _____

SET III EXERCISES

1. _____

2. _____

_____

3. _____

_____

Supplemental Exercises

1. The lengths in pages of the 10 chapters in the textbook are as follows:

1: 54	4: 62	7: 46	10: 54
2: 62	5: 82	8: 78	
3: 62	6: 74	9: 76	

a. Complete the table below to make a frequency distribution of the page lengths of the 10 chapters.

Length in pages	Tally marks	Frequency
40–49	_____	___
50–59	_____	___
60–69	_____	___
70–79	_____	___
80–89	_____	___

b. Make a histrogram illustrating this frequency distribution on the axes below.

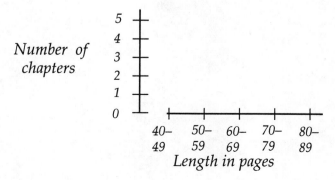

Using the Calculator

1. Use your graphing calculator to graph the following conic curve. Enter these two equations with the "y =" key.

    $$y1 = \sqrt{16 - 4x^2}$$
    $$y2 = -\sqrt{16 - 4x^2}$$

    Use standard values for the window, but change the range to y min = –7 and y max = 7. Use the graph provided below and plot the curve by tracing and locating at least four points.

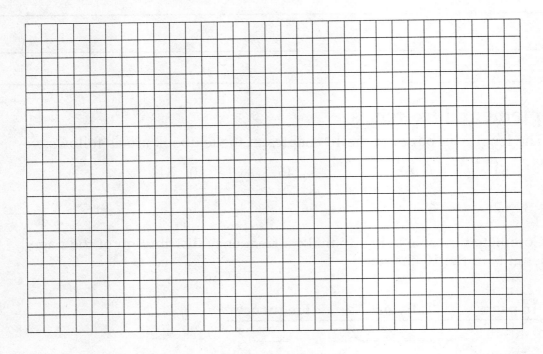

   a.   Do you use a straight edge to connect the points?   _____

   b.   What is the name of this curve?   _____

   c.   Determine its equation.   _____

SET I EXERCISES

1. _____

2. _____

3. _____   4. _____

5. _____   6. _____

7. _____   8. _____

9. _____

10. _____

11. _____

12. _____
_____

13. _____
_____

14.

Letter	Tally marks	Frequency	Letter	Tally marks	Frequency
A	_____	___	N	_____	___
B	_____	___	O	_____	___
C	_____	___	P	_____	___
D	_____	___	Q	_____	___
E	_____	___	R	_____	___
F	_____	___	S	_____	___
G	_____	___	T	_____	___
H	_____	___	U	_____	___
I	_____	___	V	_____	___
J	_____	___	W	_____	___
K	_____	___	X	_____	___
L	_____	___	Y	_____	___
M	_____	___	Z	_____	___

15. _____

16. _____   17. _____

18. _____ ; _____

19. _____

20. _____

SET II EXERCISES

1. B F C O N   A N Y K F   I X K U S   I X H U C   **A**   **N**
                                                       B   **O**
    O N G F B   C N I A C   H C N A A   K S T N I   **C**   **P**
                                                       D   **Q**
    C O N P H   B F Y K I   N K M C O   N R H W H   **E**   **R**
                                                       F   **S**
    F N A N F   H T L C O   B A W S H   L N I H F   **G**   **T**
                                                       H   **U**
    B P W K U   C H F C U   K S N B F   C O N H P   **I**   **V**
                                                       J   **W**
    N U B Y H   F T B Y C   K U L H C   C O N I N   **K**   **X**
                                                       L   **Y**
    Y B A B T   N D H C C   S N K M P   B I X H L   **M**   **Z**

2. _____

3. _____

4. _____

5. _____

SET III EXERCISES

_____

_____

Reinforcing Past Lessons

1. Make two line graphs, one for men and another for women, for the "Average Age at First Marriage" from 1890 to 1990.

Year	Men	Women
1890	26.1	22.0
1910	25.1	21.6
1930	24.3	21.3
1950	22.8	20.3
1970	23.2	20.8
1990	26.1	23.9

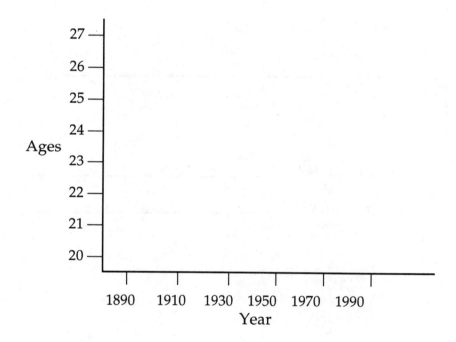

a. How would you describe the trends in words?

_____

_____

2. On a true/false test of six questions, how many different sets of answers are possible (for instance, "TTTFFF")? Without listing all possible answers, explain your answer mathematically.

_____

_____

a. What method of probability did you use to determine your answer?

_____

3. How many ways are there to scramble the letters in the word WORD?

    _____

4. If 11 horses are entered in a race, how many ways can the horses come in first, second, and third place (win, place, and show)?

    _____

5. Graph the equation $y = 3$ sine $2x$ on the axes provided.

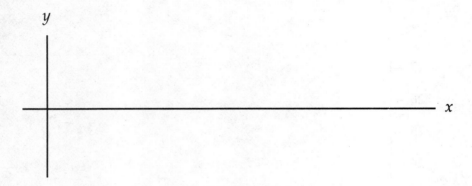

a. What is the amplitude?   _____

b. What is the frequency?   _____

SET I EXERCISES

The *mean* of a set of numbers is found by adding them and dividing the result by the number of numbers added.

The *median* of a set of numbers is the number in the middle when the numbers are arranged in order of size.

The *mode* of a set of numbers is the number that occurs most frequently, if there is such a number.

1. _____   2. _____

3. _____ 4. _____ 5. _____

6. _____

7. _____ 8. _____ 9. _____

10. _____

11. _____ 12. _____

13. _____ 14. _____

15. _____ 16. _____

17. _____ 18. _____

19. _____ 20. _____

21. _____ 22. _____

23. _____ 24. _____

25. _____

26. _____

SET II EXERCISES

1. _____   2. _____

3. _____

4. _____

5. _____

6.

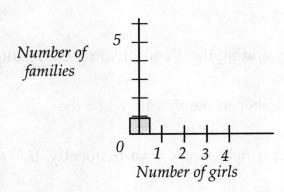

7. _____

8. _____

9.

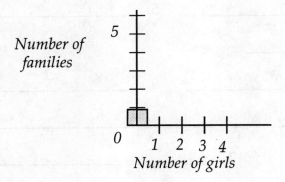

10. _____

11. _____

12. _____

13. _____

14. _____

15. _____

SET III EXERCISES

1.

Class	Amount of money in billions
1	___
2	___
3	___
4	___
5	___

2.

Class	Amount of money in billions
1	_____
2	_____
3	_____
4	_____
5	_____

3. _____

4. _____

5. _____

6. _____

Supplemental Exercises

1. On page 177 you made a histogram of the page length of the chapters in the textbook. The page lengths for the 10 chapters are:

   1: 54   4: 62   7: 46   10: 54
   2: 62   5: 82   8: 78
   3: 62   6: 74   9: 76

   a. What is the mean of this set of numbers? _____

   b. What is the median of this set of numbers? _____

   c. What (if any) is the mode of this set of numbers? _____

Reinforcing Past Lessons

1. The chart below shows the average life expectancy of a person born in the United States during the given years. Make a graph of this information.

Year	Age
1920	54.1
1940	62.9
1960	69.7
1980	73.7
1990	75.4

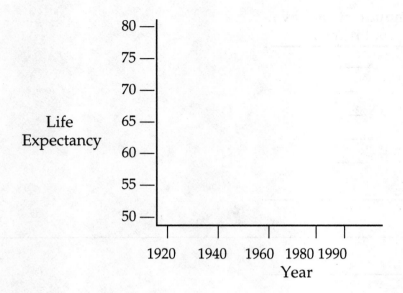

a. Using the graph, estimate the life expectancy for the year 1950.

_____

b. What type of estimation is this?   _____.

c. Using the graph, estimate the life expectancy for the year 2000.

_____

d. What type of estimation is this?   _____.

Name _____

Date _____

SET I EXERCISES

The *range* of a set of numbers is the difference between the largest and the smallest numbers in the set.

The *standard deviation* of a set of numbers is determined by finding:
1. the mean, or average of the numbers,
2. the differences between each number in the set and the mean,
3. the squares of these differences,
4. the mean of the squares, and
5. the square root of this mean.

1. _____ 2. _____

3.

Time	Difference from mean	Square of difference	Time	Difference from mean	Square of difference
114	6	36	120	__	____
116	__	____	121	__	____
116	__	____	121	__	____
119	__	____	125	__	____
120	__	____	128	__	____

4. _____ 5. _____

6. _____ 7. _____

8. _____ 9. _____

10.

Time	Difference from mean	Square of difference	Time	Difference from mean	Square of difference
104	16	256	120	__	____
116	__	____	121	__	____
116	__	____	121	__	____
119	__	____	125	__	____
120	__	____	138	__	____

11. _____ 12. _____

13. _____ 14. _____

15. _____ 16. _____

17. _____ 18. _____

19. _____ 20. _____

21. _____

SET II EXERCISES

1. _____ 2. _____

3. _____ 4. _____ 5. _____

6. _____ 7. _____ 8. _____

9. _____ 10. _____ 11. _____

12. _____

13. _____ 14. _____

15. _____ 16. _____ 17. _____

18. _____ 19. _____ 20. _____

21. _____ 22. _____ 23. _____

24. _____ 25. _____ 26. _____

27. _____ 28. _____

29. _____

SET III EXERCISES

1. Newly minted ____ After 5 years ____ After 10 years ____

2. _____

3. _____

Name _____

Date _____

SET I EXERCISES

1.

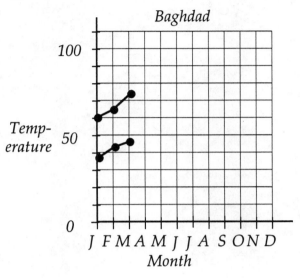

2.

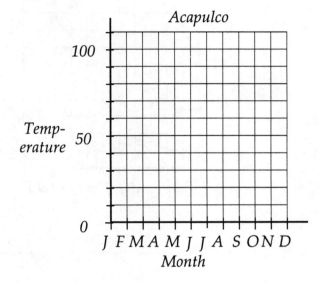

3.

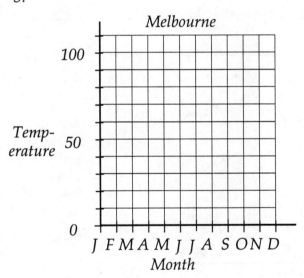

4. _____

5. _____

6. _____

7.

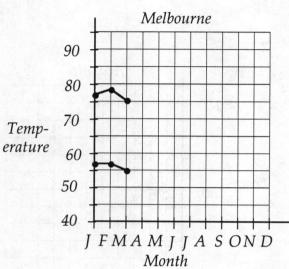

8. _____

9. _____

10. _____   11. _____

12.   96,   94,   93,   ____, ____, ____, ____, ____, ____, ____,

13.

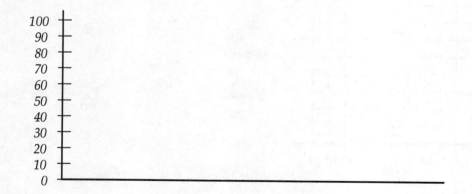

14. _____

15. _____   16. _____ ; _____

17. _____

18. _____

_____

SET II EXERCISES

1. _____

2. _____

3. _____

4. _____

5. _____

6. _____ ; _____

7. _____

8. _____   9. _____

10.

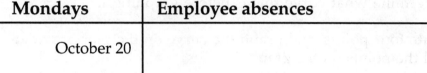

Mondays	Employee absences
October 20	
October 27	
November 3	
November 10	

0   150   300   450   600   750

11. _____

12. _____

13. _____

14. _____   15. _____

16.

SET III EXERCISES

1. _____

   _____

2. _____

   _____

3. _____

   _____

4. _____

   _____

Using the Calculator

1. Use a graphing calculator to graph the following curve: $y = 4x^2 - 5$

a. Can you determine what curve this is before graphing it? _____

b. Trace to locate four points, and graph the curve on the plot provided
    below. Label the points on the graph.

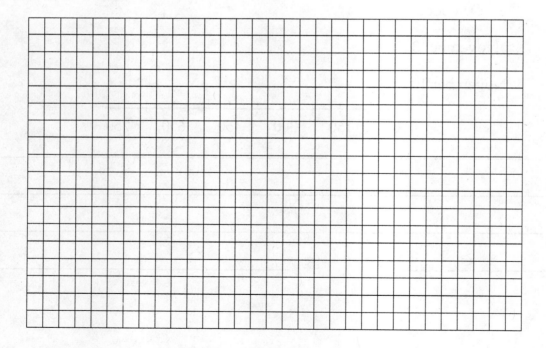

c. Is this curve symmetrical? If so, what types of symmetry does it have?

_____

LESSON 6

Name _____

Date _____

SET I EXERCISES

1. _____

   _____

2. _____

3. _____

   _____

4. _____

   _____

5. _____

6. _____

   _____

7. _____   8. _____

9. _____

10. _____

11. _____

12. _____

13. _____

SET II EXERCISES

1. _____

2. _____

3. _____   4. _____   5. _____

6. _____

7. _____

8. _____

9. _____

10. _____

11. _____

12. _____

13. _____

14. _____
_____

SET III EXERCISES

1. _____

2. _____

3. _____

4. _____

5. _____

6. _____

CHAPTER 9
Summary and Review

Name _____

Date _____

SET I EXERCISES

1. _____   2. _____

3. _____   4. _____

5. _____   6. _____

7. _____   8. _____

9. _____
   _____

10. _____   11. _____

12. _____   13. _____

14. _____   15. _____   16. _____

17. _____   18. _____   19. _____

20. _____ ; _____
    _____

SET II EXERCISES

1. _____ ; _____
   _____

2. _____

3. _____   4. _____

5. _____ ; _____
   _____

6. _____

7. _____

8.

Time in mins.	Tally marks	Frequency	Time in mins.	Tally marks	Frequency
71–80	_____	___	111–120	_____	___
81–90	_____	___	121–130	_____	___
91–100	_____	___	131–140	_____	___
101–110	_____	___			

9.

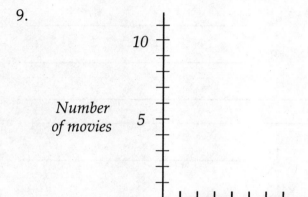

10. _____   11. _____

12. _____ ; _____

_____

13. _____

14. _____

15. _____

16. _____ ; _____

SET III EXERCISES

1. _____   2. _____   3. _____

4. _____

5. _____   6. _____

7. _____   8. _____

9. _____

Reinforcing Past Lessons

1. The Pythagorean theorem states that the sum of the squares of the legs of a right triangle is equal to the square of the hypotenuse. If the legs are 5 feet and 12 feet, what is the length of the hypotenuse?

    _____

2. Use base 2 logarithms to solve this problem without multiplying.
    512 × 131,072   _____

3. Identify the solid with the numeration 3-3-3-3 and explain the numeration.

    _____

4. What is the probability that involves two, and *only* two, possible outcomes?

    _____

5. You will use the page lengths of the chapters in the textbook again for this exercise. Use the space at the bottom of the page to work out your answers.

1: 54	4: 62	7: 46	10: 54
2: 62	5: 82	8: 78	
3: 62	6: 74	9: 76	

a. What is the range of this set of numbers?

    _____

b. What is the standard deviation of this set of numbers?

    _____

6. Is the picture graph below representing the page lengths of the longest and shortest chapters in the textbook fair? Why or why not.

_____ , _____

_____

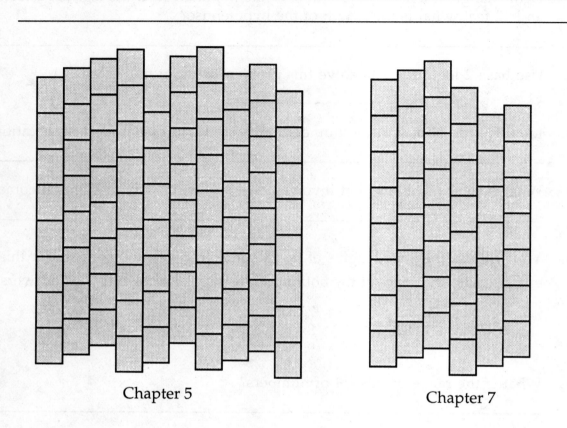

Chapter 5   Chapter 7

CHAPTER **10**
LESSON 1

Name _____

Date _____

SET I EXERCISES

A *simple closed curve* is a curve on which it is possible to start at any point and move continuously around the curve, passing through every other point exactly once before returning to the starting point.

Figures are *topologically equivalent* if they can be stretched and bent into the same shape without connecting or disconnecting any points.

1. _____   2. _____   3. _____

4. _____

5. _____

6. _____   7. _____   8. _____

9. _____   10. _____

11. _____

12. _____

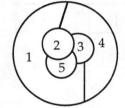

13. _____

14. _____ ; _____

15. _____

SET II EXERCISES

1. _____

_____

2. _____   3. _____

4. _____

5. _____

6. _____

7. _____

8. _____

9. _____

10.

W	G	E

A	B	C

11. _____   12. _____

13. _____

14. _____

15.

```
   A    C

 B       B
     A
 C
```

16.

```
 A       B



 B       A
```

17.

```
 A           B
    C  D
    B  A
 D           C
```

If you think any of these three will not work in the way specified, explain why not:

_____
_____

SET III EXERCISES

1. _____

2. _____ ; _____
_____

3. _____

Supplemental Exercises

Match the lettered drawings in the first column to those numbered drawings in the second are topologically equivalent. There may be one, more than one, or no matches for each numbered drawing.

(Think of the lines as pieces of rubber bands that can be stretched and shrunk.)

A)
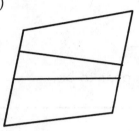

1)
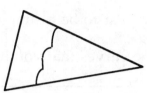

B)
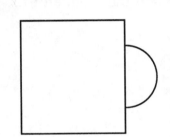

2)
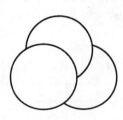

C)
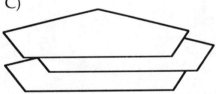

3)
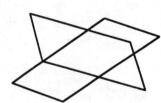

D)
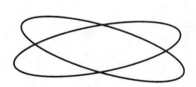

4)
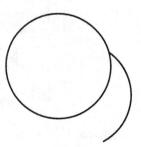

E)
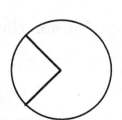

1) _____

2) _____

3) _____

4) _____

Using the Calculator

1. Use your graphing calculator to graph the following equations:

    $$y = \sqrt{(-x^2 + 16)}$$
    $$y = -\sqrt{(-x^2 + 16)}$$

    Use the standard window of {x: –10, 10} and {y: –10, 10}.

    a. Does this appear to be a simple closed curve? _____

    b. Which of the curves that you have studied does this appear to be?

    _____

    c. Copy your graph onto the grid below. Trace points on the calculator to insure that your graph is correct.

    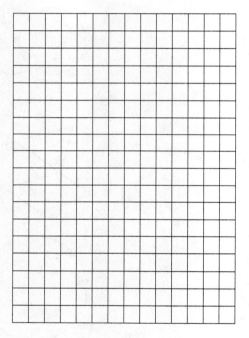

    d. Does the curve look the same on the graph above as it did on the calculator? If not, explain. _____

    e. Use "zoom standard" to get a different window on your calculator. Now what curve does it appear to be? _____

    f. Explain this difference in appearance.

    _____
    _____

SET I EXERCISES

1. _____ 2. _____

3. _____ 4. _____ 5. _____ 6. _____

7.

Network	Network letter	Number of even vertices	Number of odd vertices	Can the network be traveled?
	A	4	0	Yes
	B	—	—	—
	C	—	—	—
	D	—	—	—
	E	—	—	—
	F	—	—	—
	G	—	—	—
	H	—	—	—
	I	—	—	—

8. _____

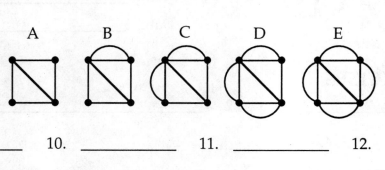

A   B   C   D   E

9. _____   10. _____   11. _____   12. _____

(For Exercises 13 through 15, draw arrows on each line to show that you have traveled it. Place a large open circle on the beginning vertex, and a closed circle on the last vertex.)

13.
Route map A

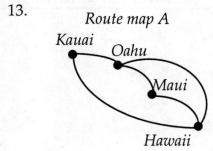

14.
Route map B

15.
Route map C

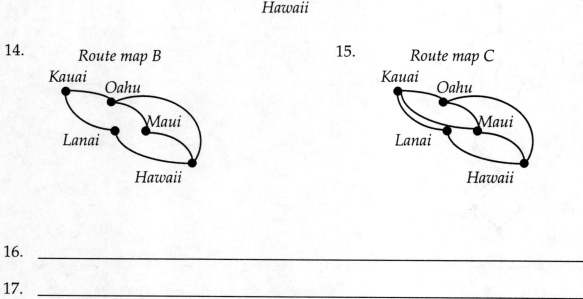

16. _____

17. _____

18. _____   19. _____   20. _____

SET II EXERCISES

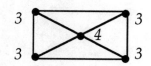

1. _____

2. _____   3. _____

(Use colored pencils to trace networks 4 through 13, and mark with arrows as you proceed.)

4.

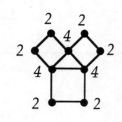

5.

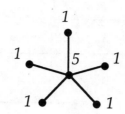

6.

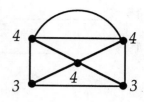

7.

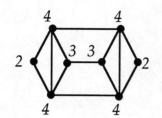

8.

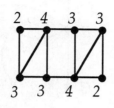

9.

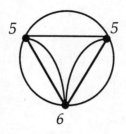

10.

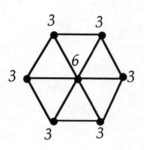

11.

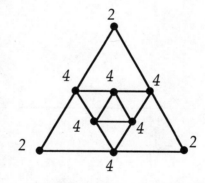

12.

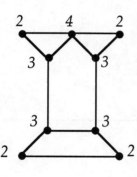

13.

14.

Network	Number of even vertices	Number of odd vertices	Can the network be traveled?
4	9	0	Yes
5	___	___	___
6	___	___	___
7	___	___	___
8	___	___	___
9	___	___	___
10	___	___	___
11	___	___	___
12	___	___	___
13	___	___	___

15. _____   16. _____   17. _____   18. _____

19. _____

20.

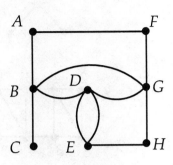

21. _____   22. _____

SET III EXERCISES

1. Draw your network in the space provided.

2. _____ ; _____

3. _____

CHAPTER 10
LESSON 3

Name _____

Date _____

SET I EXERCISES

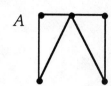
A   B   C

1. _____
_____

2. _____   3. _____   4. _____

5. _____   6. _____   7. _____

8. _____

9. _____

10. _____

11. _____   13.

12. _____

14. _____   15. _____ ; _____   16. _____ ; _____

17. _____

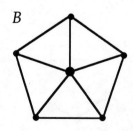   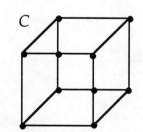   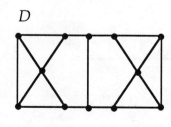
B   C   D

19.

Network	Number of odd vertices	Fewest number of trips
A	4	2
B	___	___
C	___	___
D	___	___

20. _____

_____

A

B
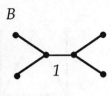

C
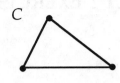

D
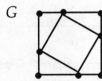

E

F
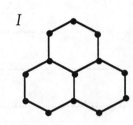

G

H
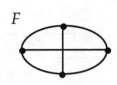

I

22.

Network	Number of regions	Number of vertices	Number of edges
A	3	2	3
B	1	6	5
C	___	___	___
D	___	___	___
E	___	___	___
F	___	___	___
G	___	___	___
H	___	___	___
I	___	___	___

22. _____

23. Regions: _____

Vertices: _____

Edges: _____

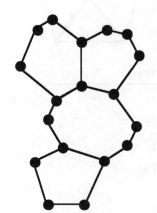

24. _____

SET II EXERCISES

1. Draw your networks in the space provided.

2. _____

3. _____ ; _____

4. _____ 5. _____ 6. _____ 7. _____

8. _____

9. _____

10. _____

Map A   *Map B*

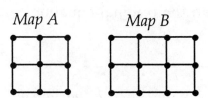

11. _____ ; _____

12.

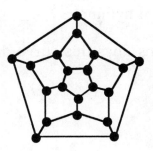

Using the Calculator

1. Graph the following equation on a graphing calculator: $y = 2x^2 - 7$.

a. Use the trace function to determine the values of "y" when "x" has the given values (round to the nearest hundredth).

x	-2	-1	0	1	2	3
y	.49	__	__	__	__	__

b.   Use these values to graph the curve on the graph below.

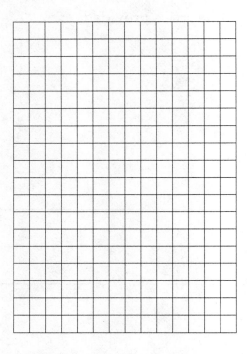

c.   Now use the same *x*-values and determine the *y*-values by substituting the *x*-values into the original equation.

x	−2	−1	0	1	2	3
y	__	__	__	__	__	__

d.   What is the name of this curve?   _____

e.   Does it have line symmetry?   _____

f.   Does it have rotational symmetry?   _____

g.   Into how many regions does this curve divide the plane?   _____

SET III EXERCISES

1.

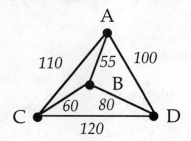

2.   _____   Draw any possible Hamiltion paths in the space provided.

3.   _____

SET I EXERCISES

1. _____ 2. _____ 3. _____ 4. _____

5. _____

6. _____

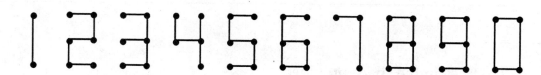

7.

Digit	Number of vertices	Number of edges	Is it a tree?
1	2	1	Yes
2	____	____	____
3	____	____	____
4	____	____	____
5	____	____	____
6	____	____	____
7	____	____	____
8	____	____	____
9	____	____	____
0	____	____	____

8. _____ 9. _____

10. Draw a tree (if possible) in the space provided.

11. _____ ; _____

12. _____ ; _____

13. _____ 14. _____ 15. _____ 16. _____

17. _____ 18. _____ 19. _____ 20. _____

SET II EXERCISES

1. _____   2. _____   3. _____   4. _____

5. _____   6. _____   7. _____   8. _____

9. _____   10. _____   11. _____   12. _____

13. _____   14. _____

15. _____ ; _____

16. _____

17. Draw the trees in the space provided.

SET III EXERCISES

1. _____   2. _____

3.

4.

5. _____

	Number of carbon atoms	Number of hydrogen atoms
6. **Hydrocarbon**		
Methane	1	4
Ethane	____	____
Propane	____	____
Butane	____	____
Pentane	____	____

7. _____   8. _____

Supplemental Exercises

1. The drawing below is a representation of the waterways of Indianapolis.

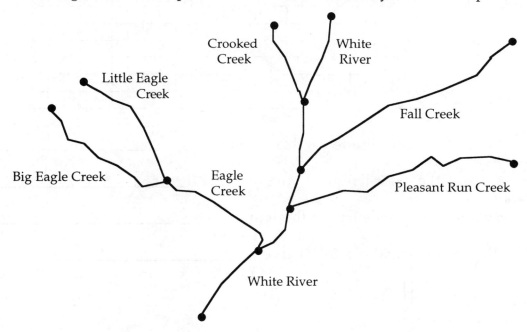

a. Does this illustrate a tree network?   _____

b. How many vertices does it have?   _____

c. How many edges does it have?   _____

d. What is its diameter?   _____

e. List one of the possible pathways along the diameter by listing the waterways along the way.   _____

_____

Using the Calculator

1. Graph the following equation on a graphing calculator: $y = 7 \sin 1.5x$.
    The calculator should be in the radian mode. Use the standard screen.

    a. Copy the graph on the grid below by tracing points for location of the curve.

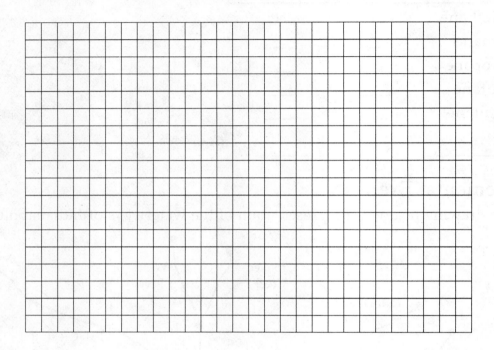

    b. What is the amplitude of this curve?   _____

    c. What is the wavelength of this curve?   _____

    d. What is the frequency of this curve?   _____

    e. What is the name of this curve?   _____

Name _____

Date _____

SET I EXERCISES

Use the graph paper provided for the Moebius strip experiments in this chapter.

1. _____ 2. _____

3. _____ 4. _____

5. _____ 6. _____

7. _____ 8. _____ 9. _____

10. _____

11. _____

12. _____ 13. _____

14. _____ ; _____

15. _____

16. _____ 17. _____

18. _____ 19. _____

20. _____

_____

SET II EXERCISES

1. _____ 2. _____ 3. _____

4. _____

5. _____

6. _____

7. _____ 8. _____

9. _____

_____

10. _____   11. _____

12. _____

13. _____

_____

14. _____

15. _____

SET III EXERCISES

1. _____

2. _____

SET I EXERCISES

For Exercises 1, 2, and 3, draw in the figure(s) in the space provided.

1.

2.

3.

●

4. _____   5. _____   6. _____

7. _____

8. _____ ; _____

9. _____   10. _____

11. _____   12. _____

13. _____   14. _____

15. _____

16. _____ ; _____

SET II EXERCISES

1. _____   2. _____   3. _____

● 4. _____

5.   6.   7.

8.   9.

10.

Network	Number of even regions	Number of odd regions	Does the network have a path?
5	2	0	Yes
6	___	___	___
7	___	___	___
8	___	___	___
9	___	___	___

11. _____

_____

12. _____ ; _____

_____

SET III EXERCISES

1. _____ ; _____

_____

2. _____ ; _____

3. _____

APPENDIX A

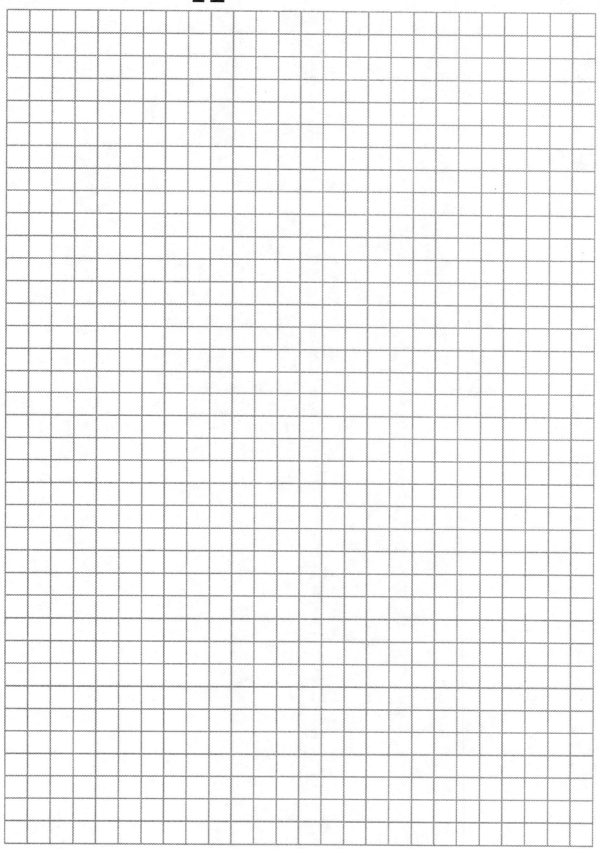

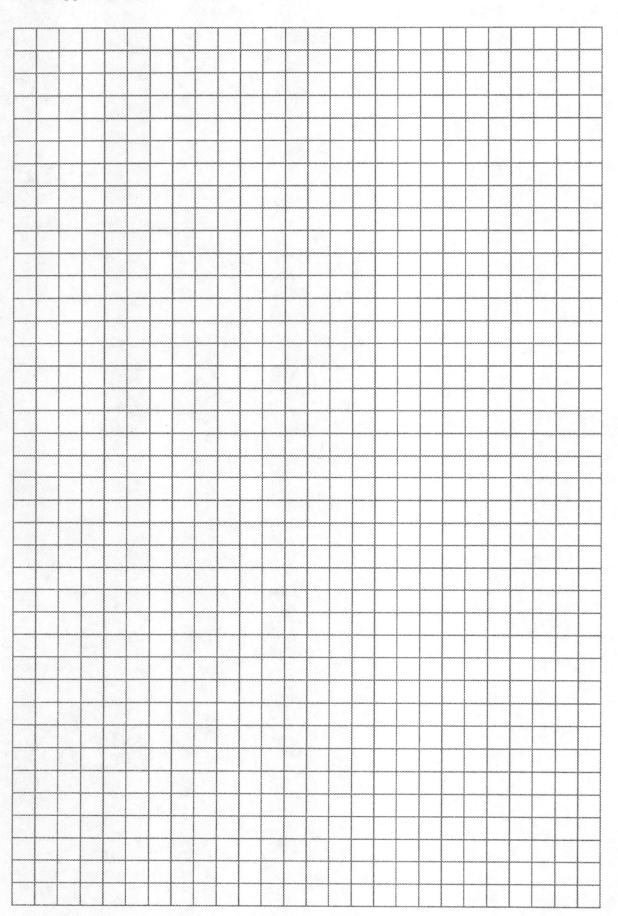

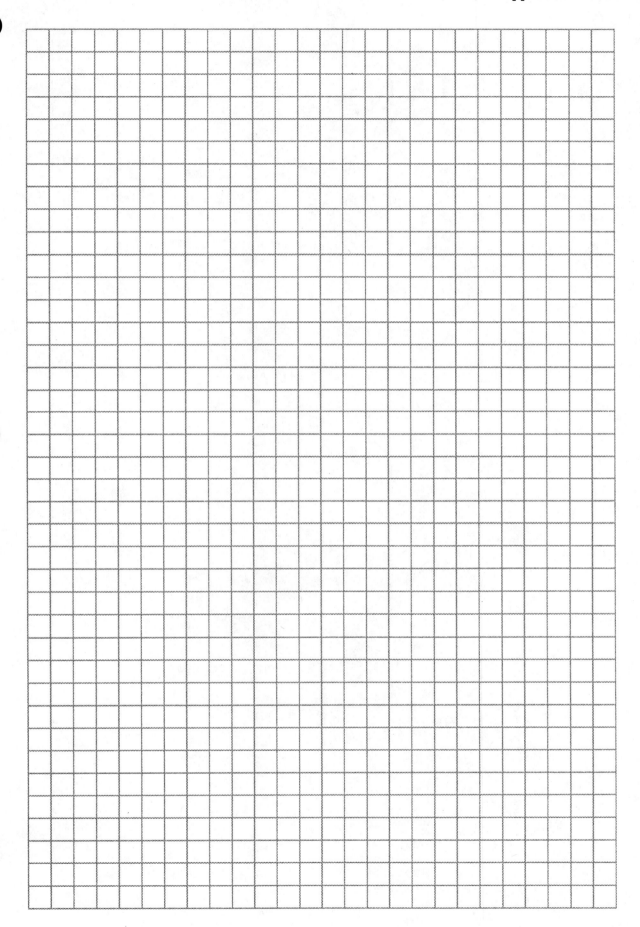

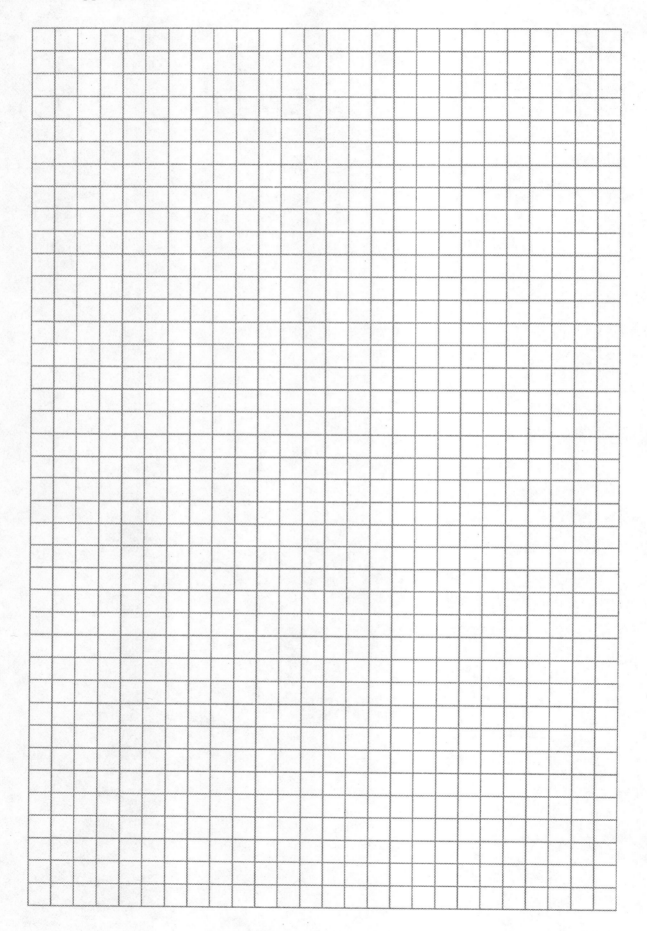

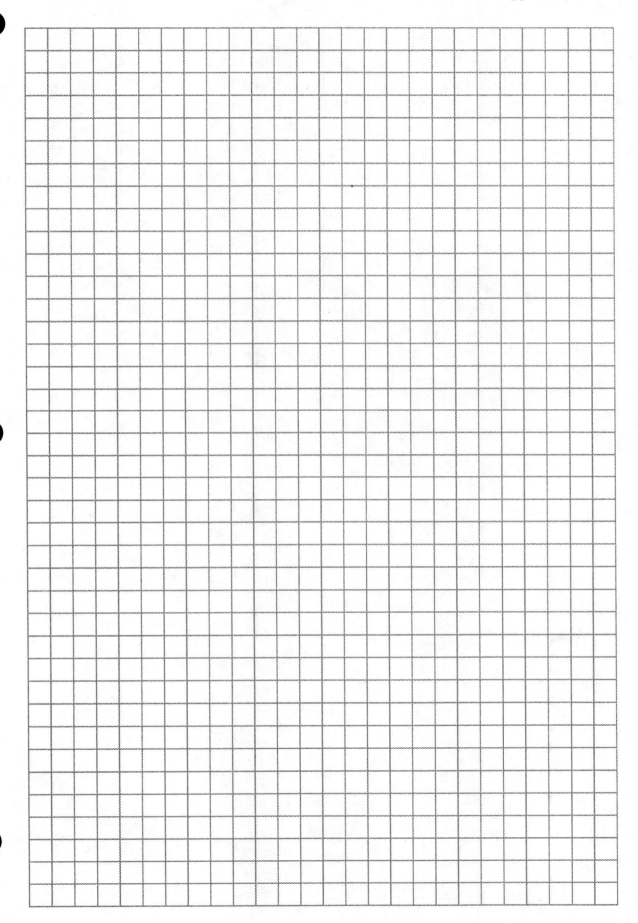

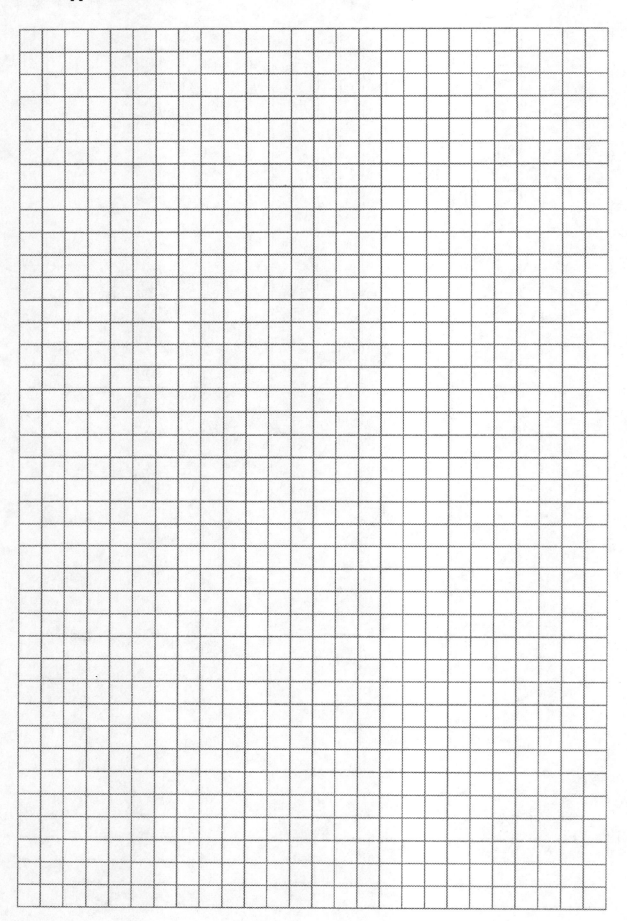

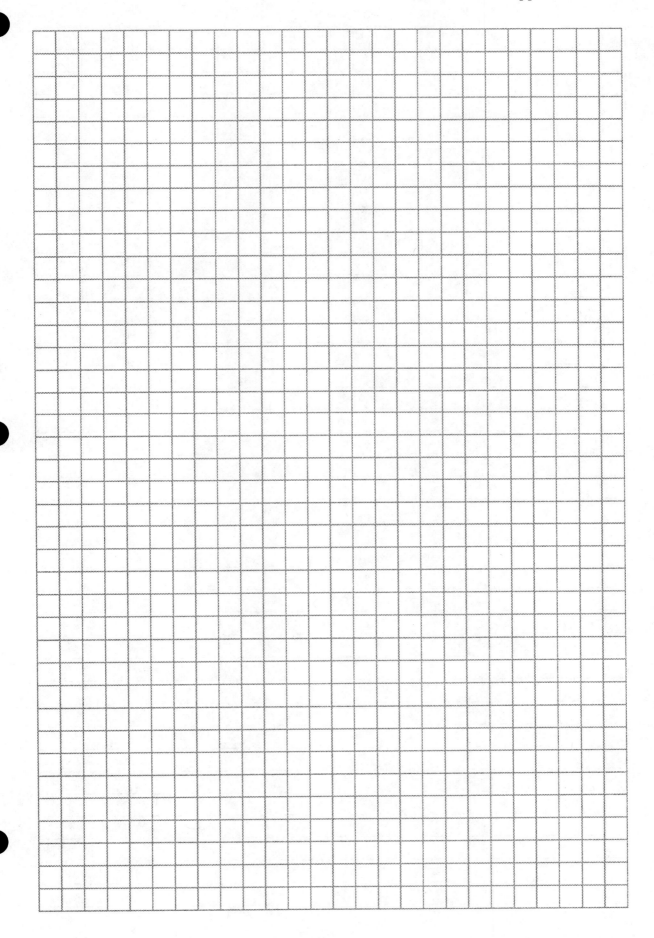

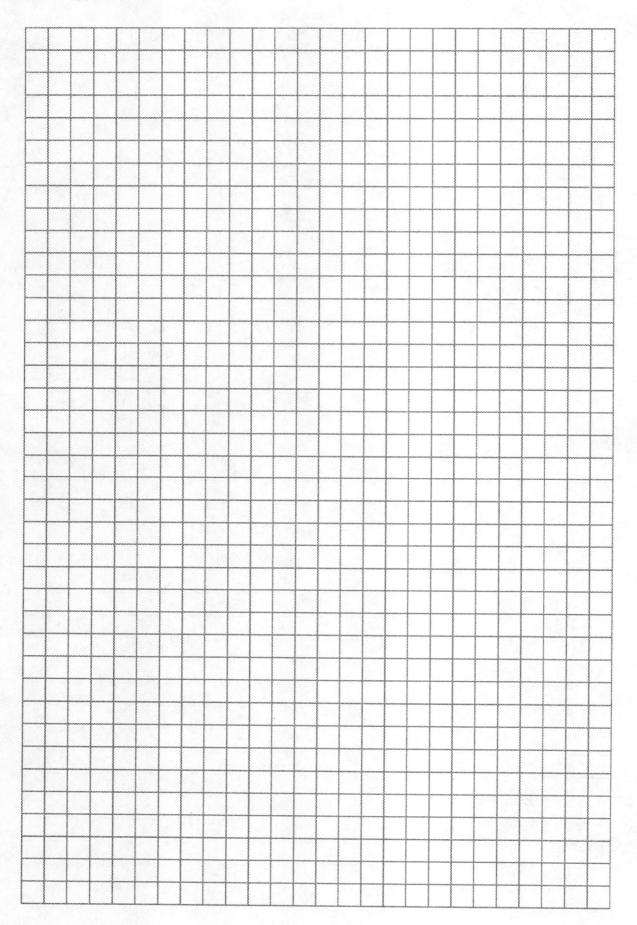

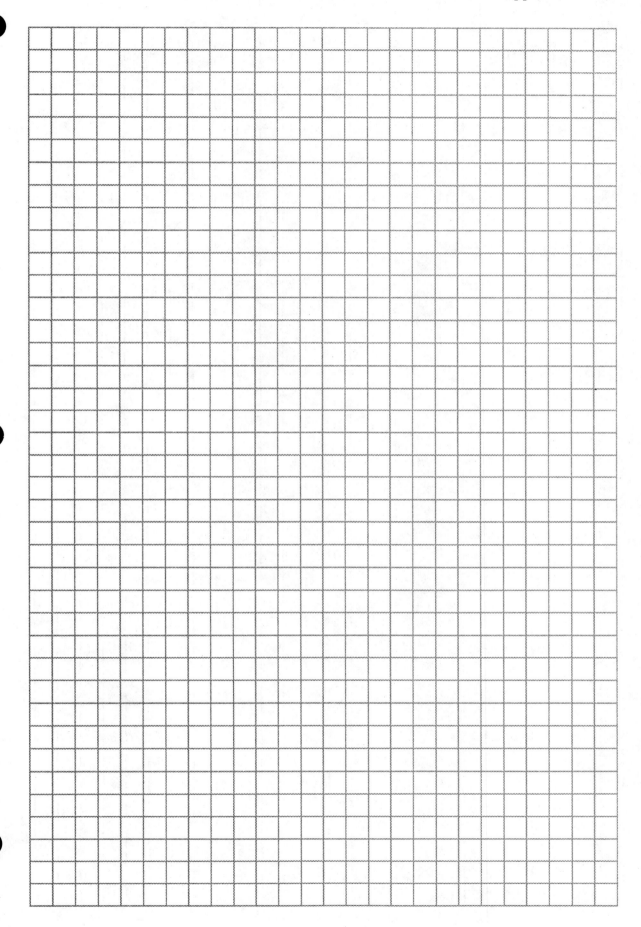

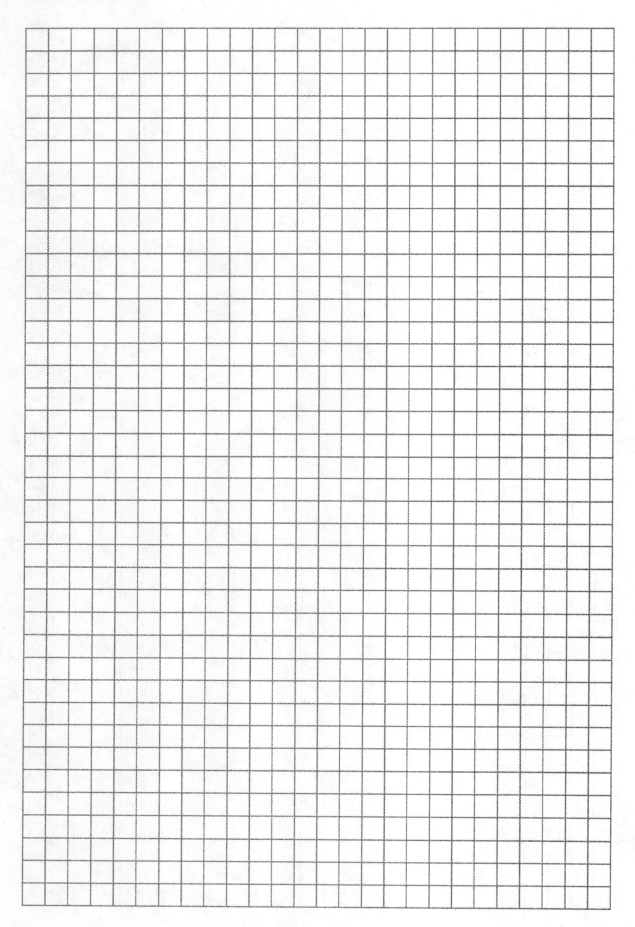

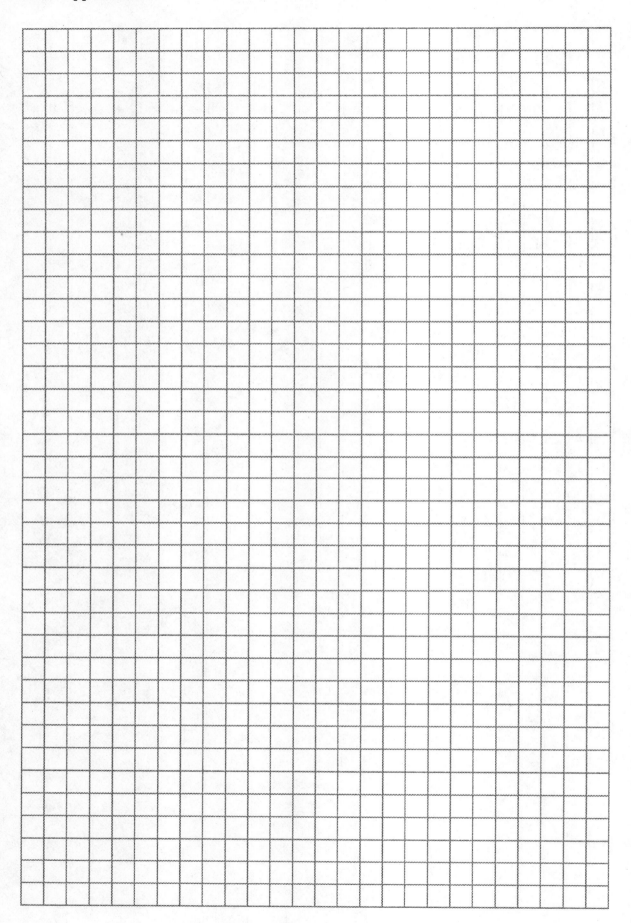

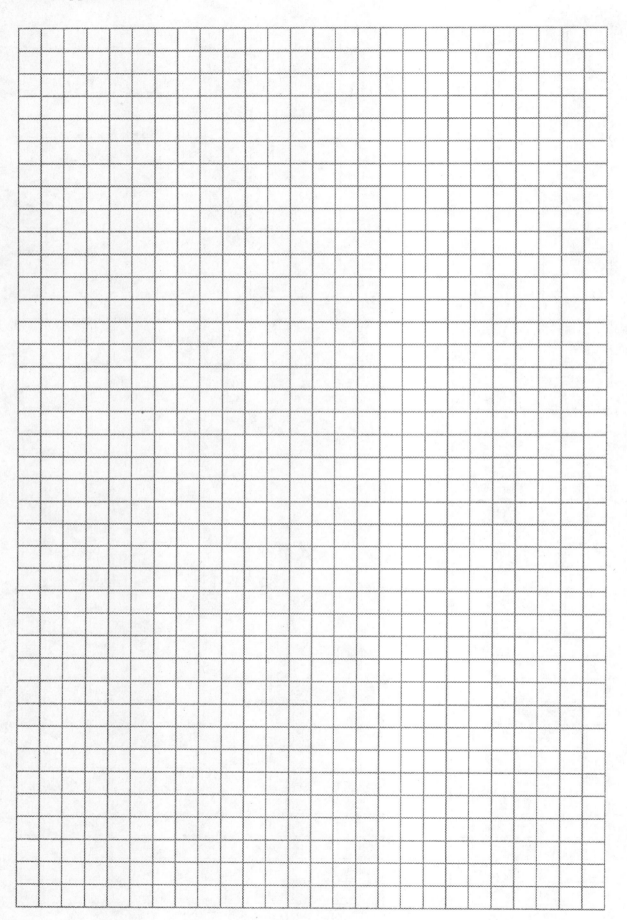

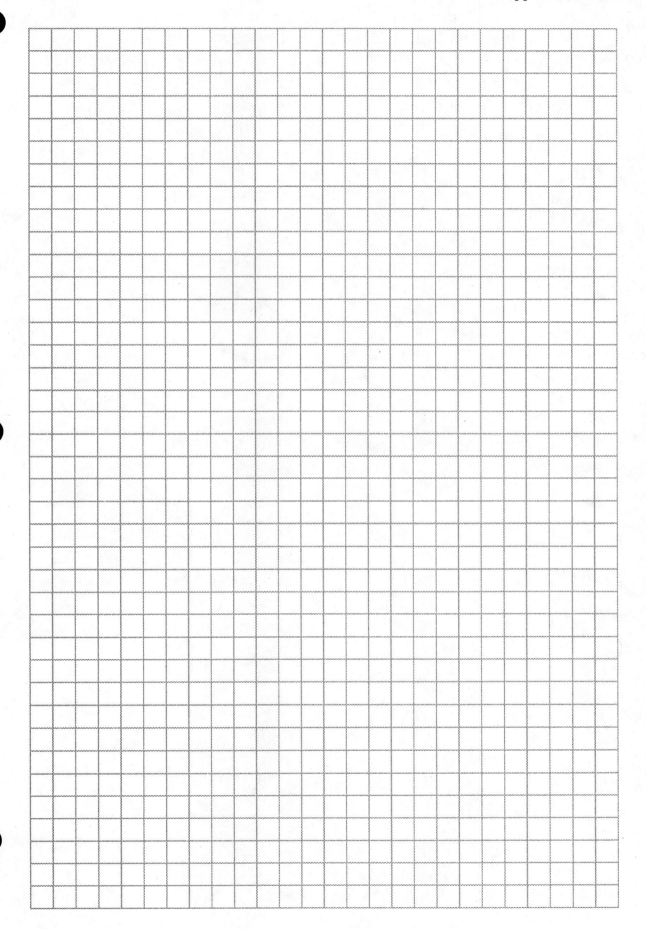

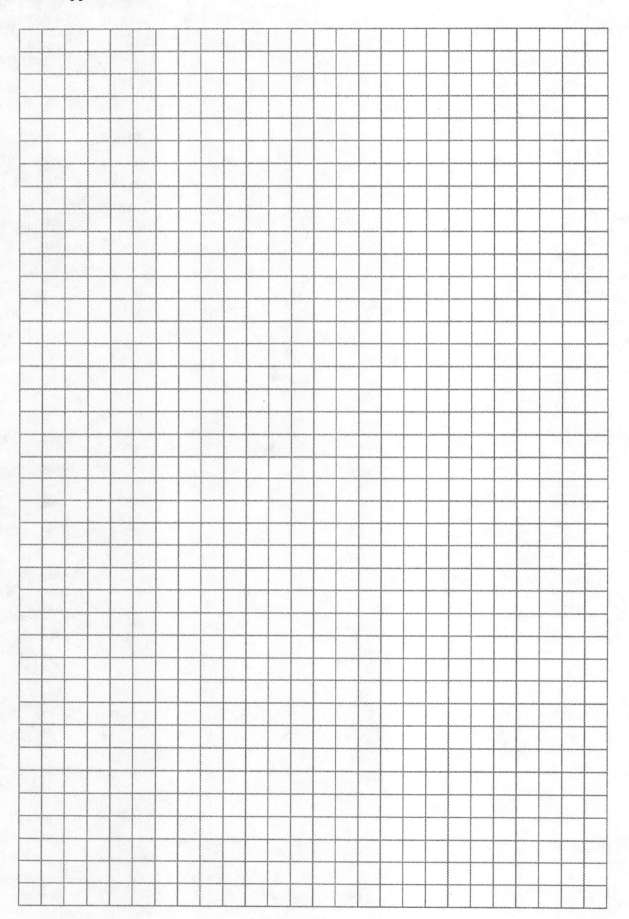

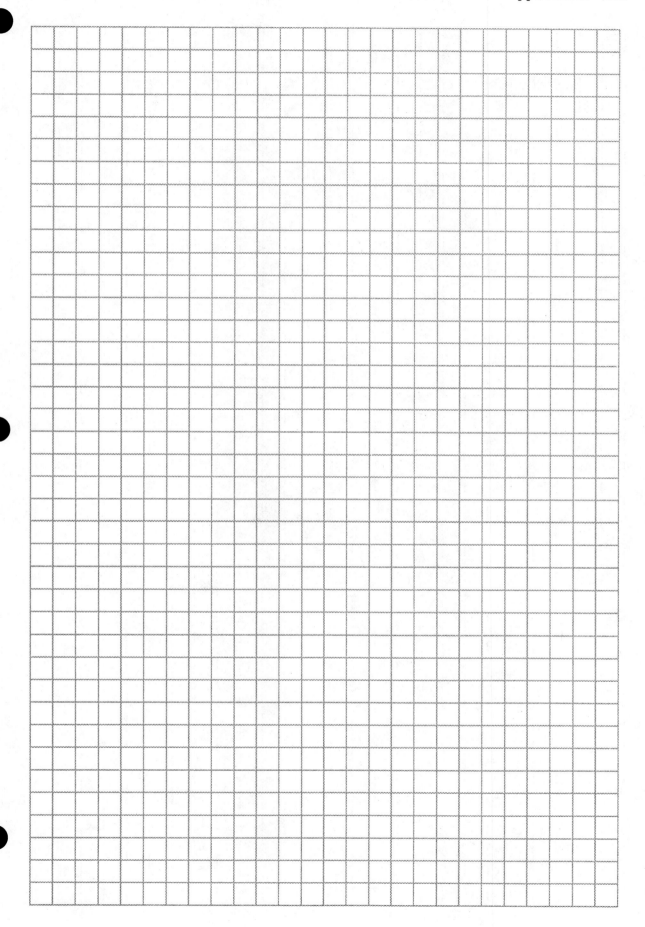

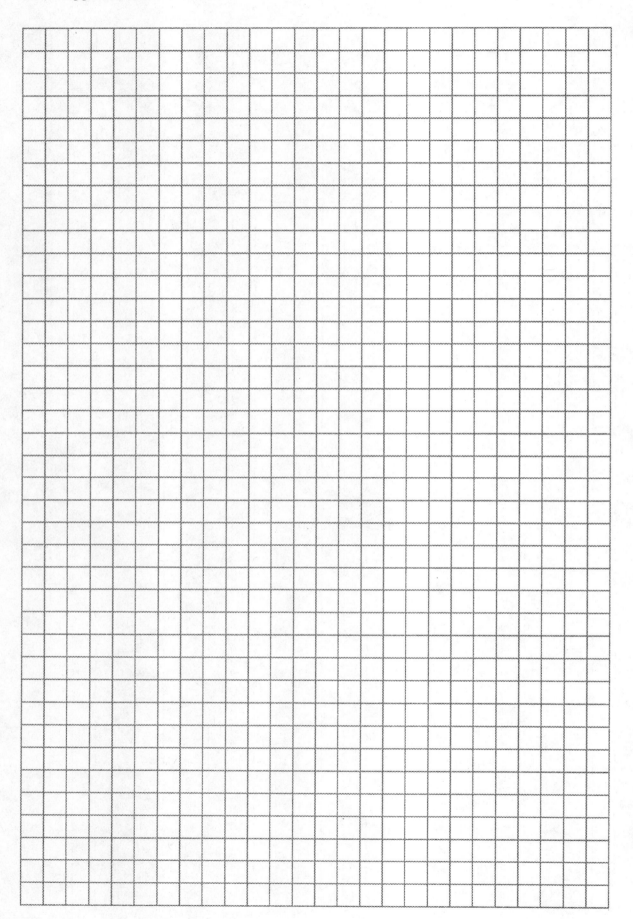

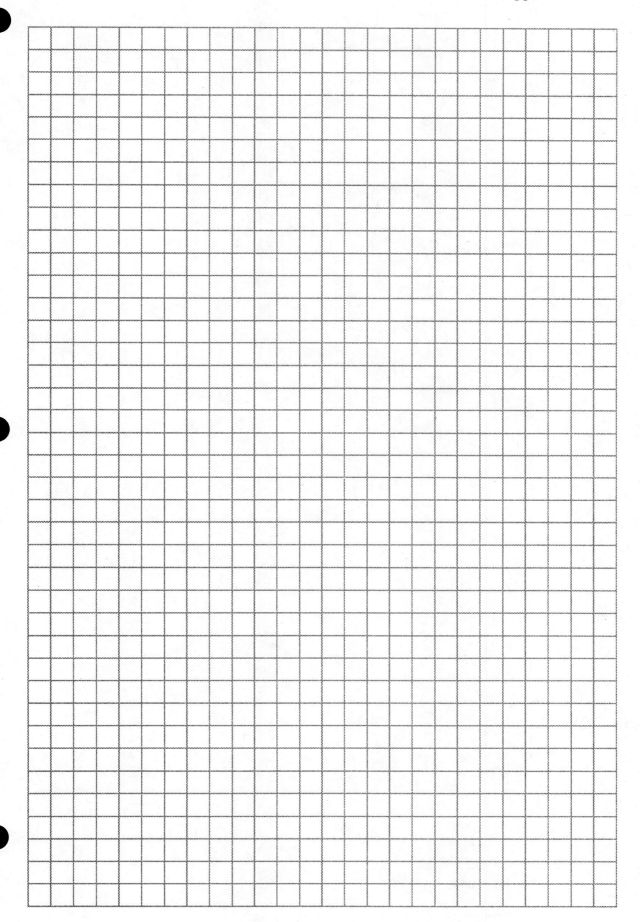

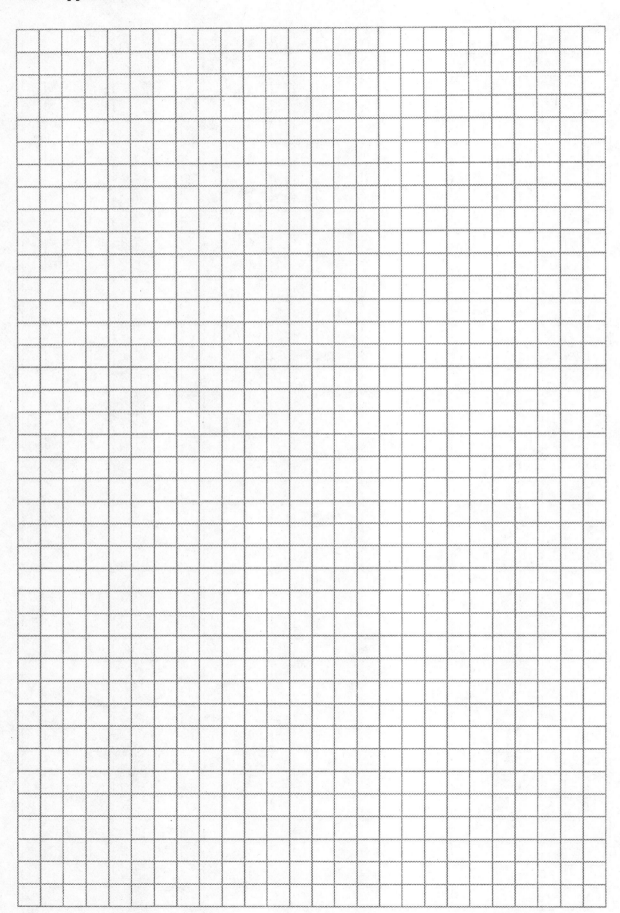

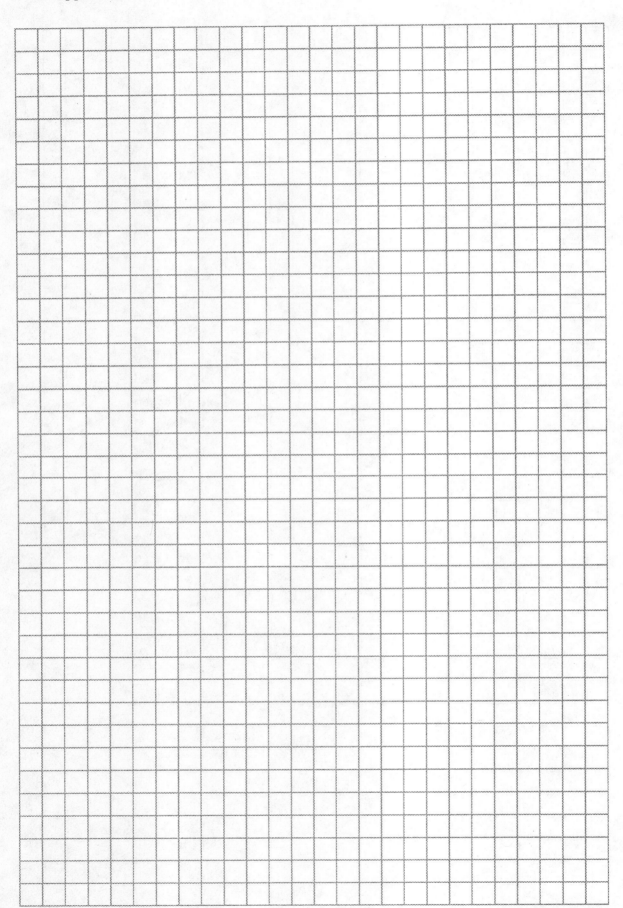

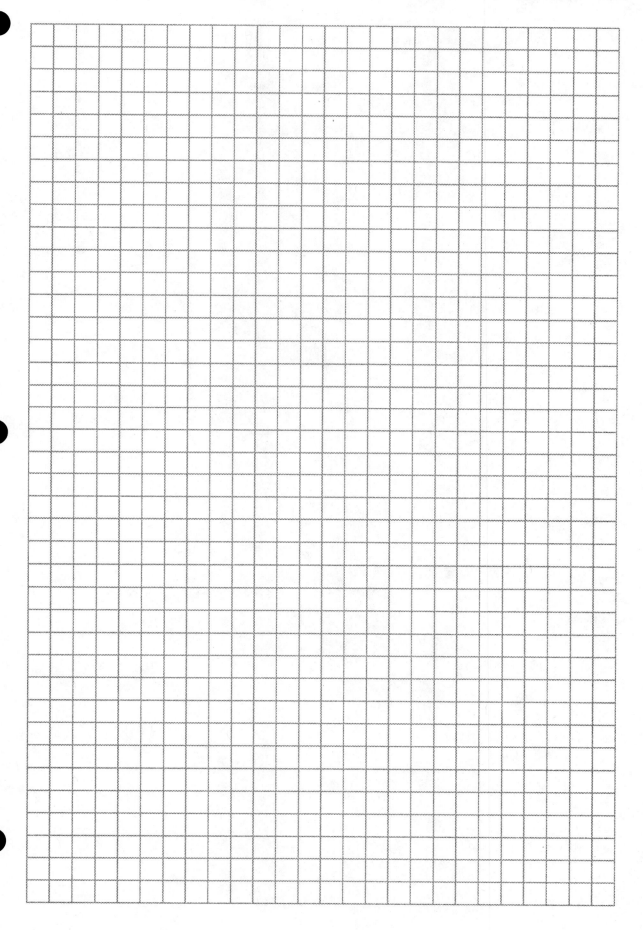

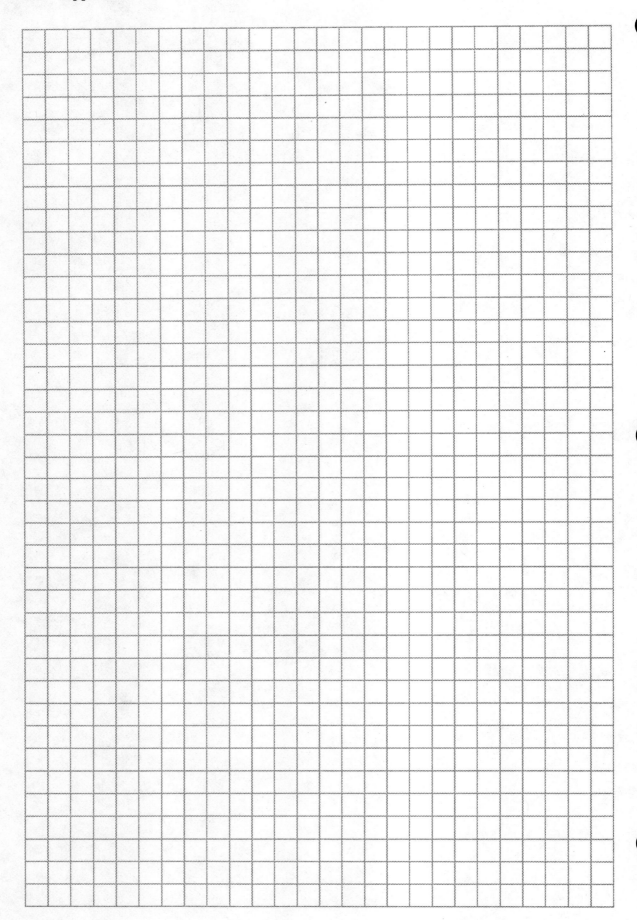

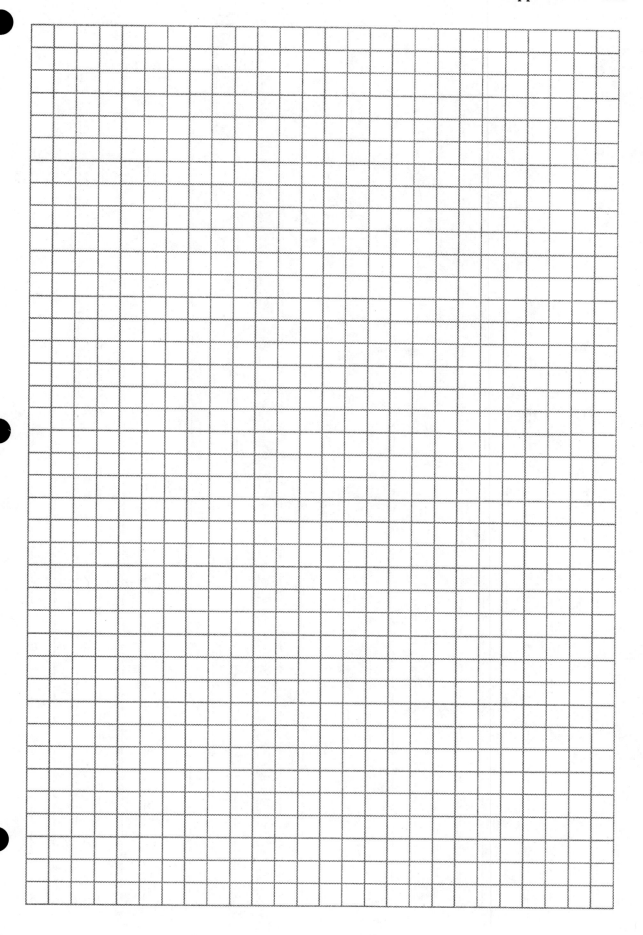

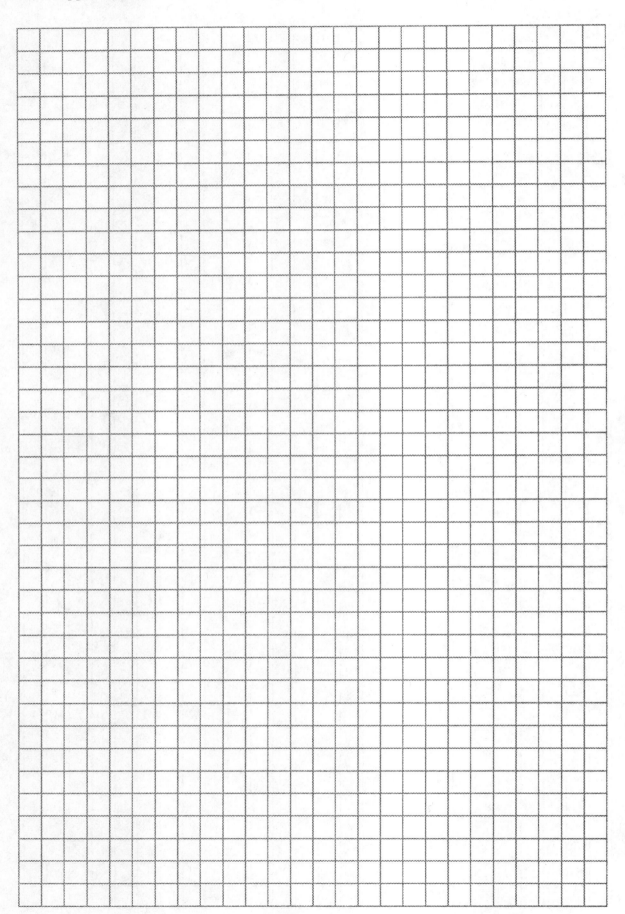

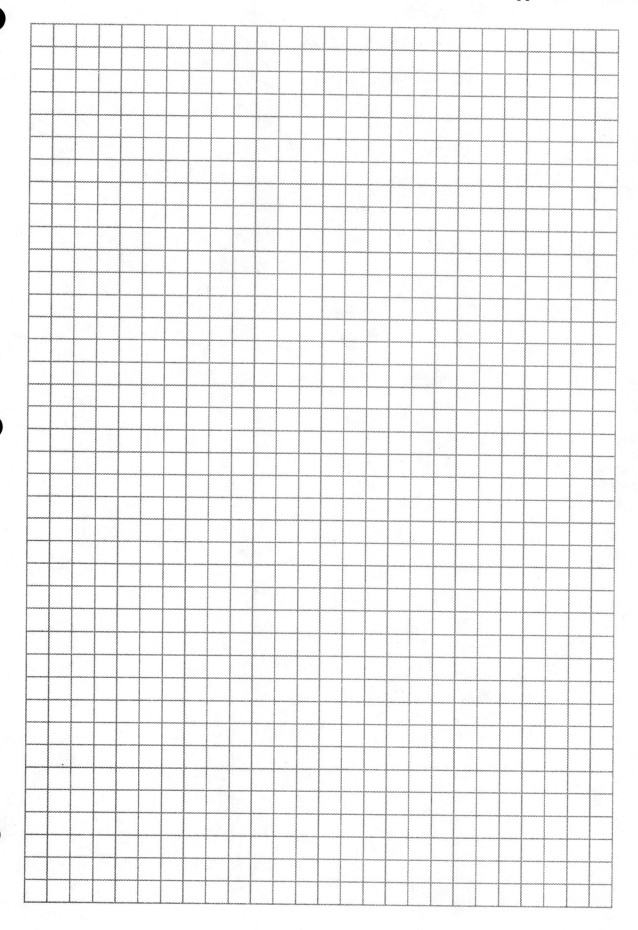

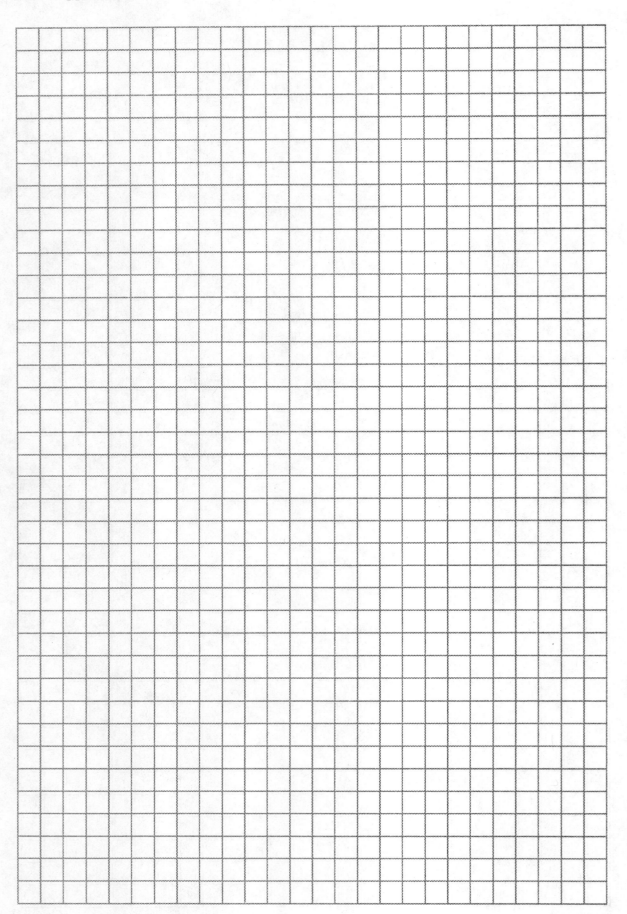

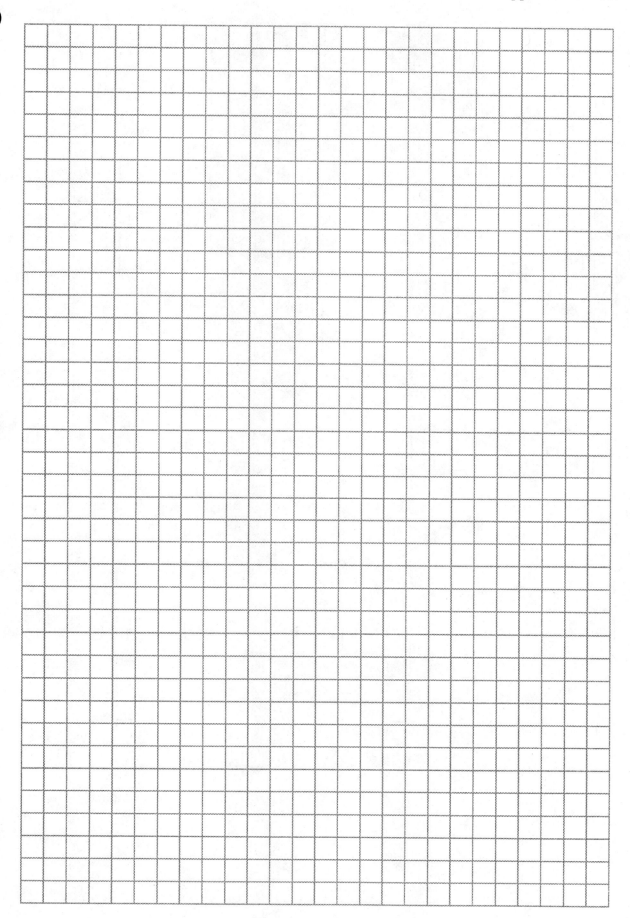

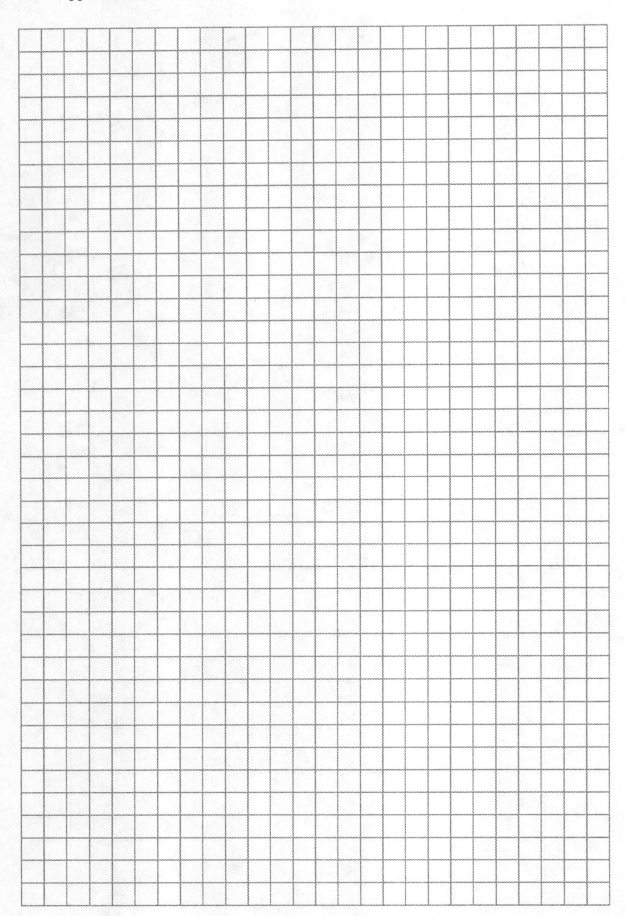

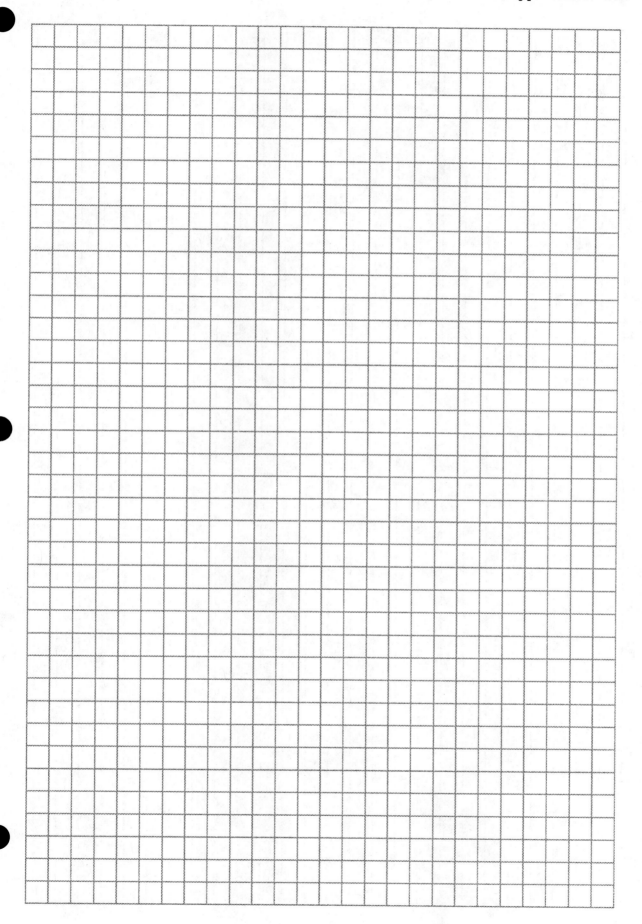

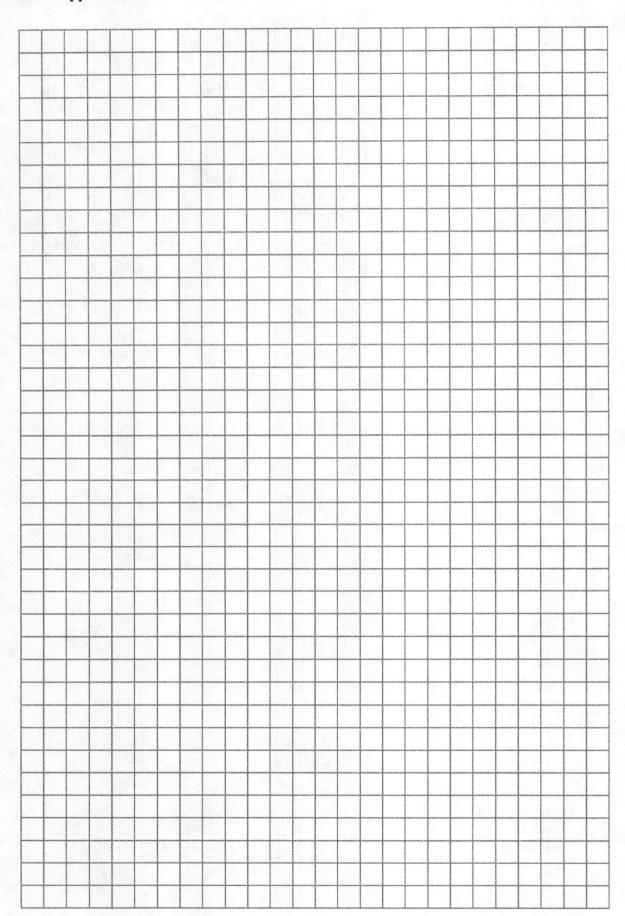

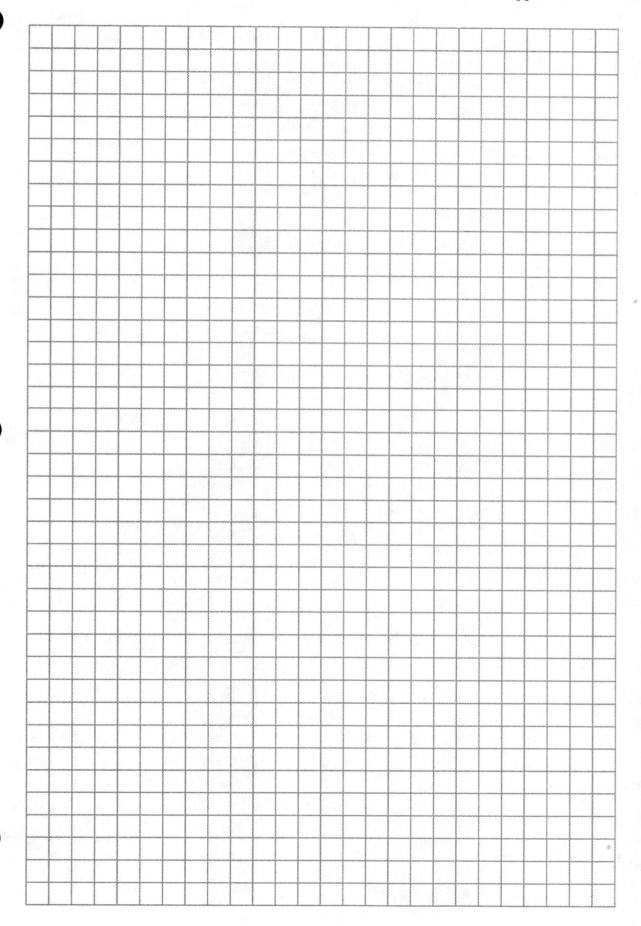

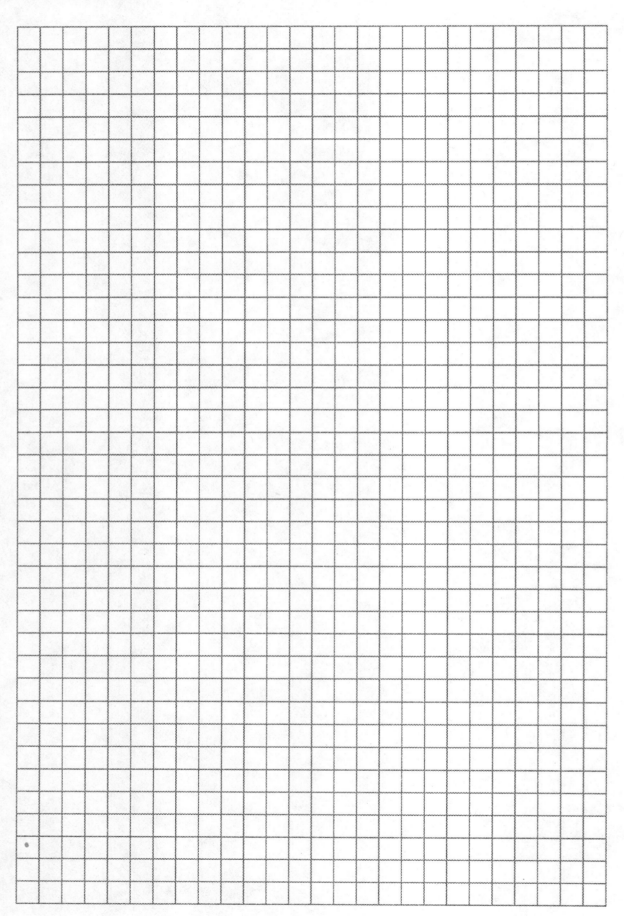

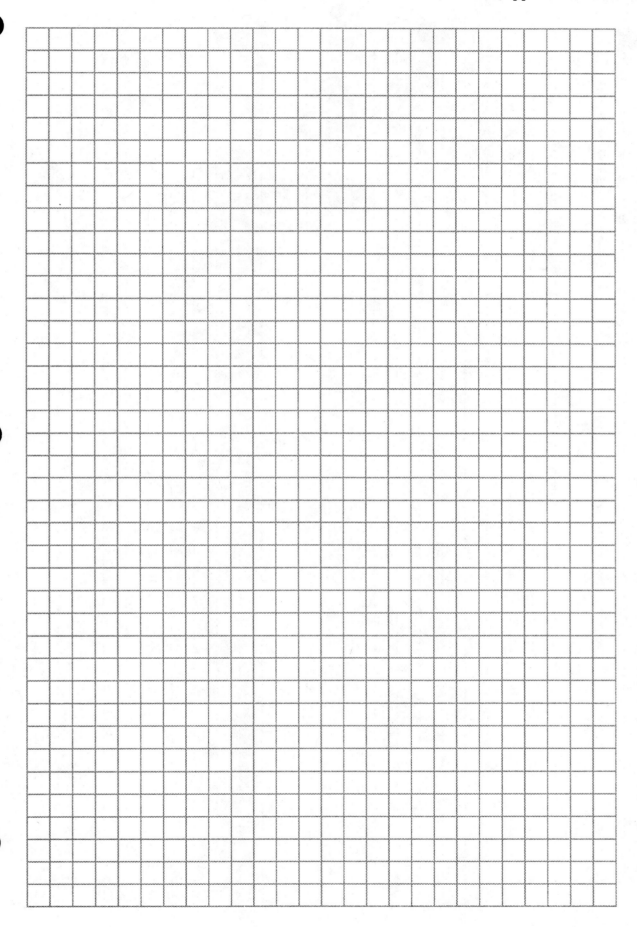

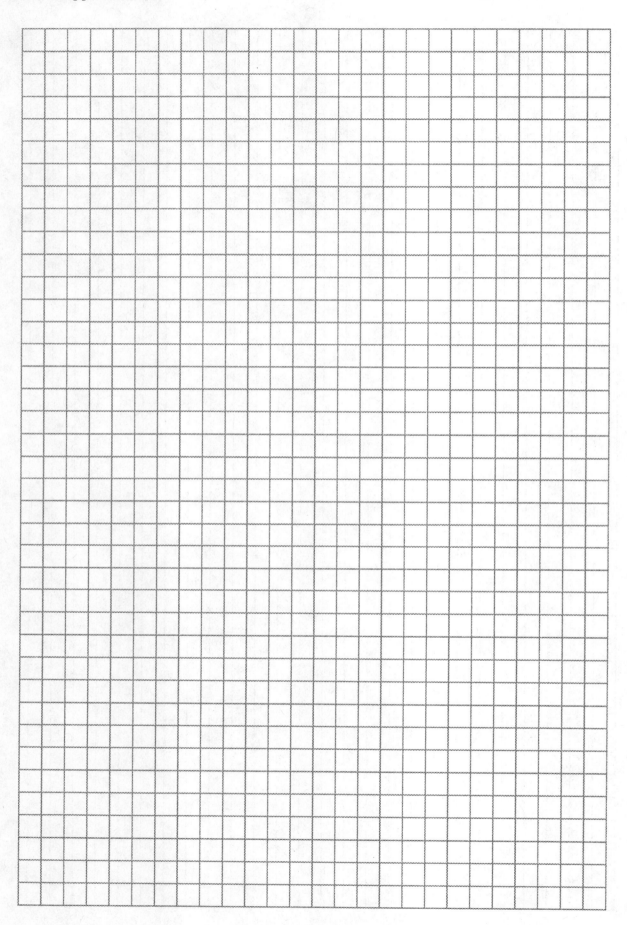

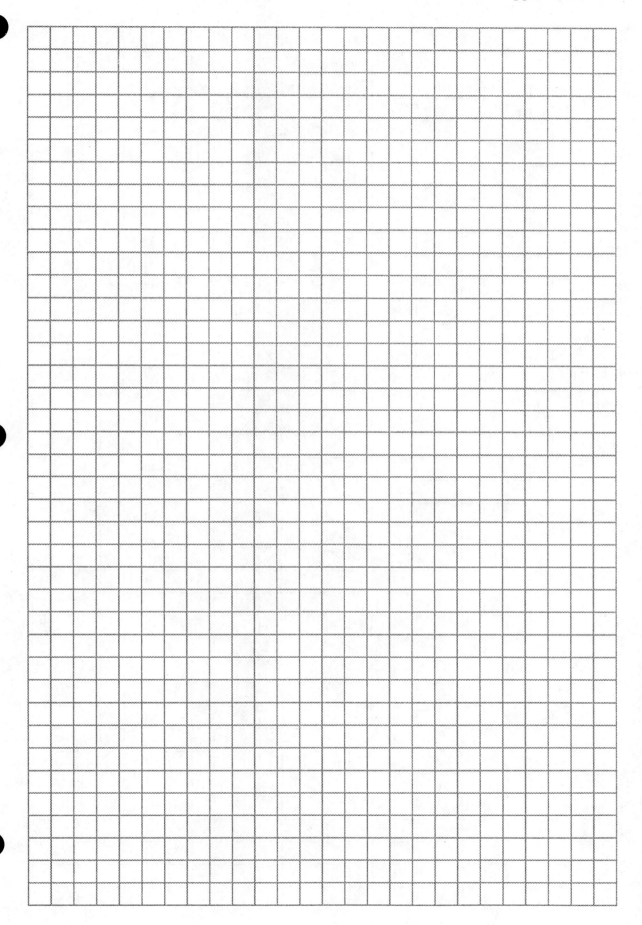

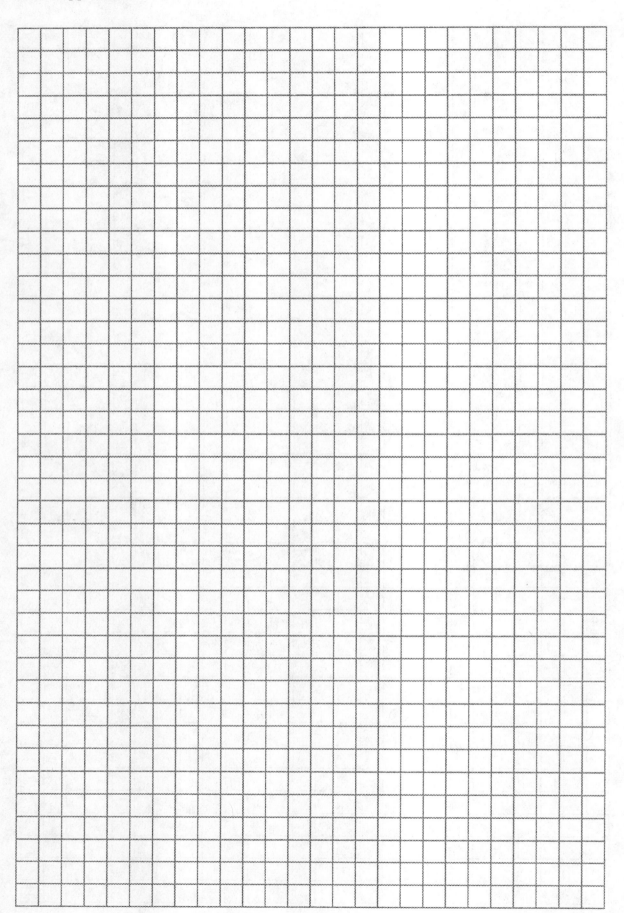

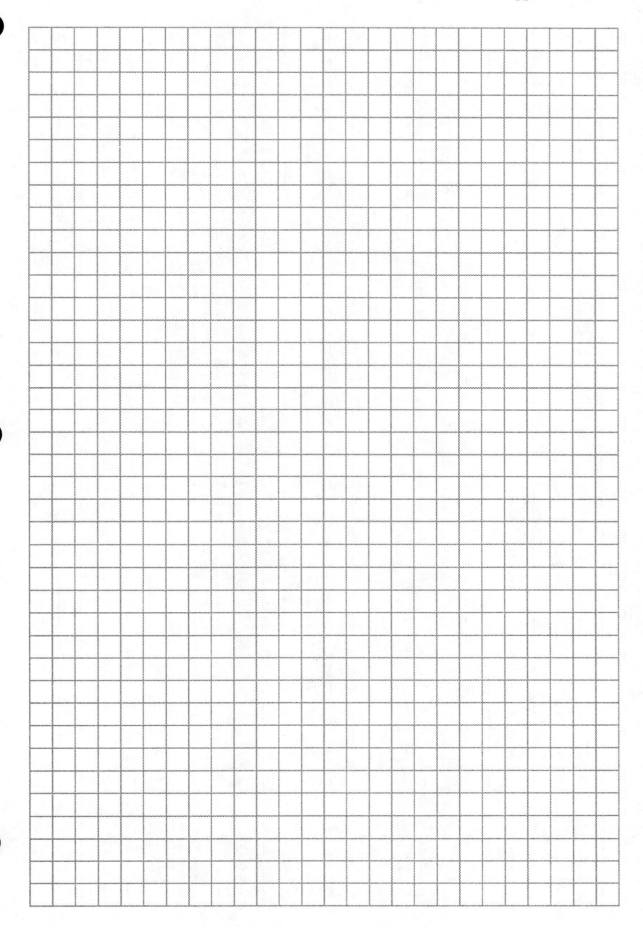

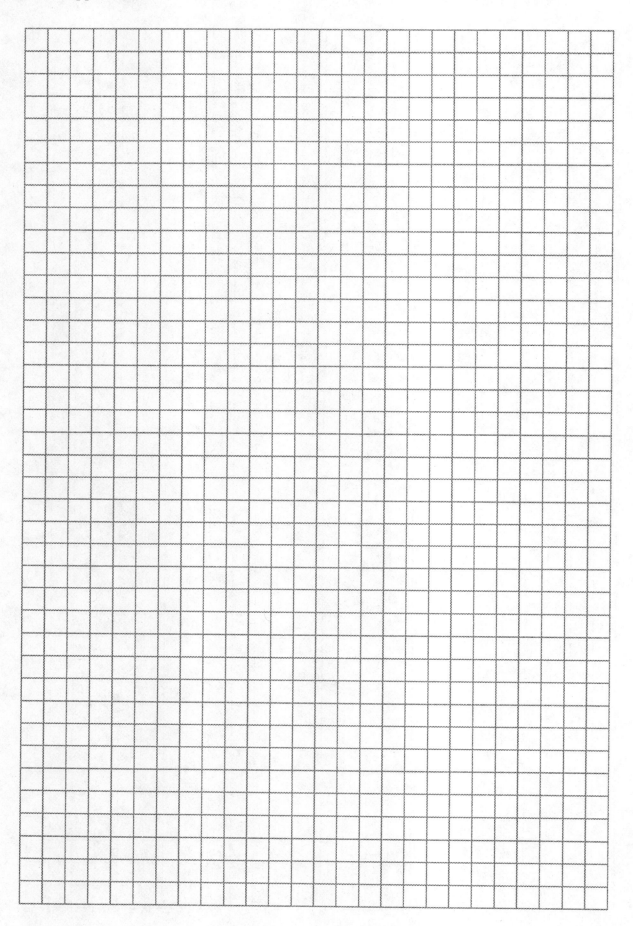

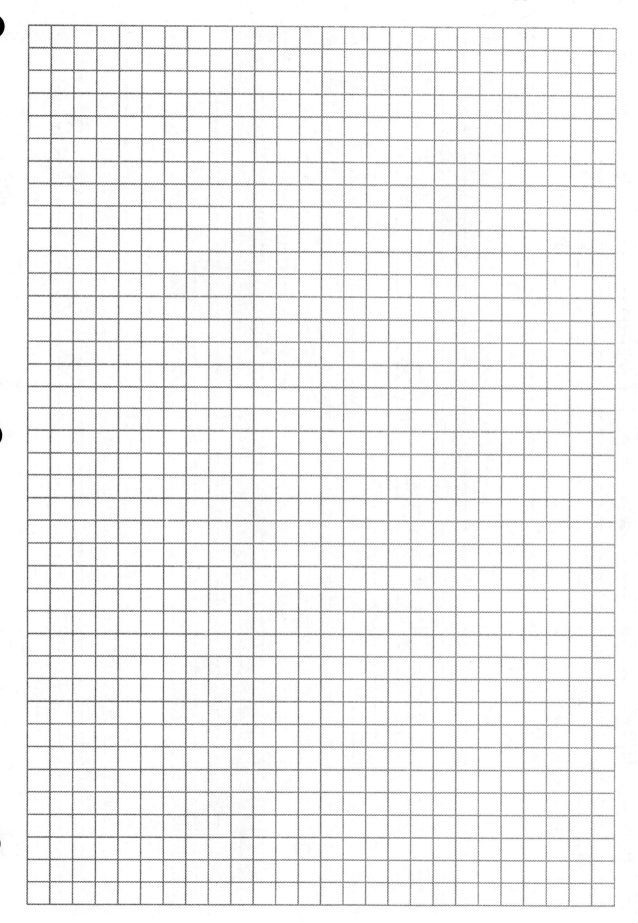

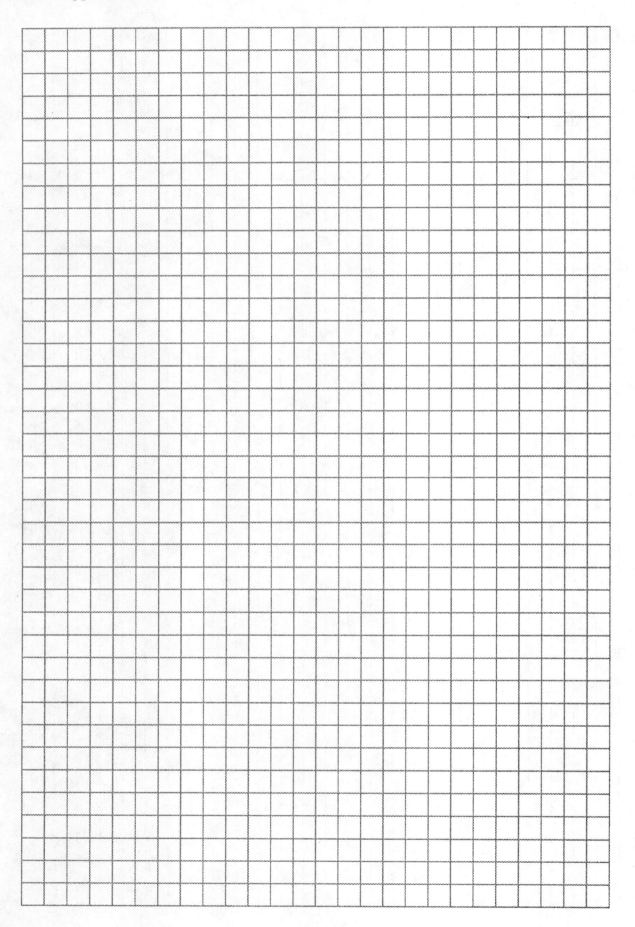

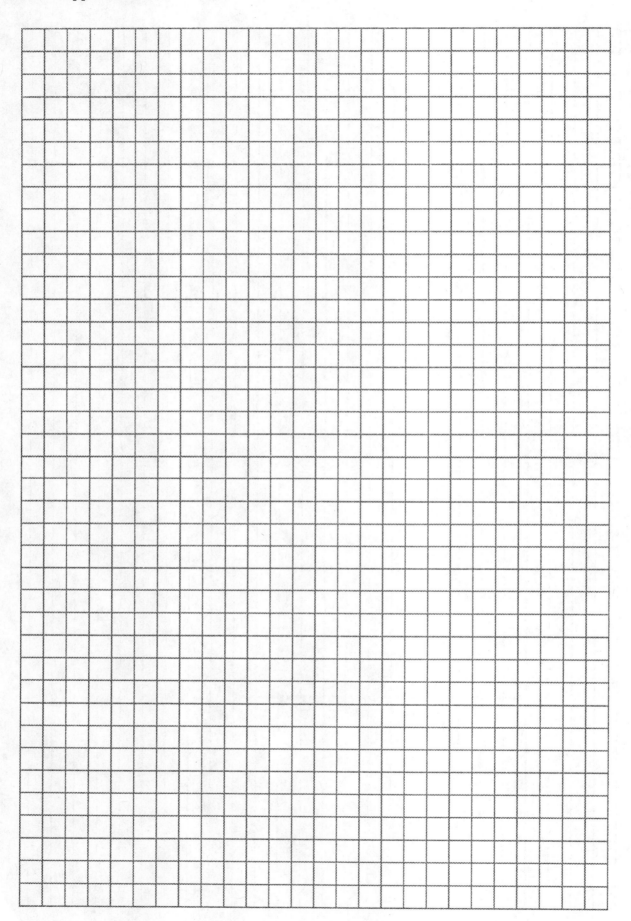

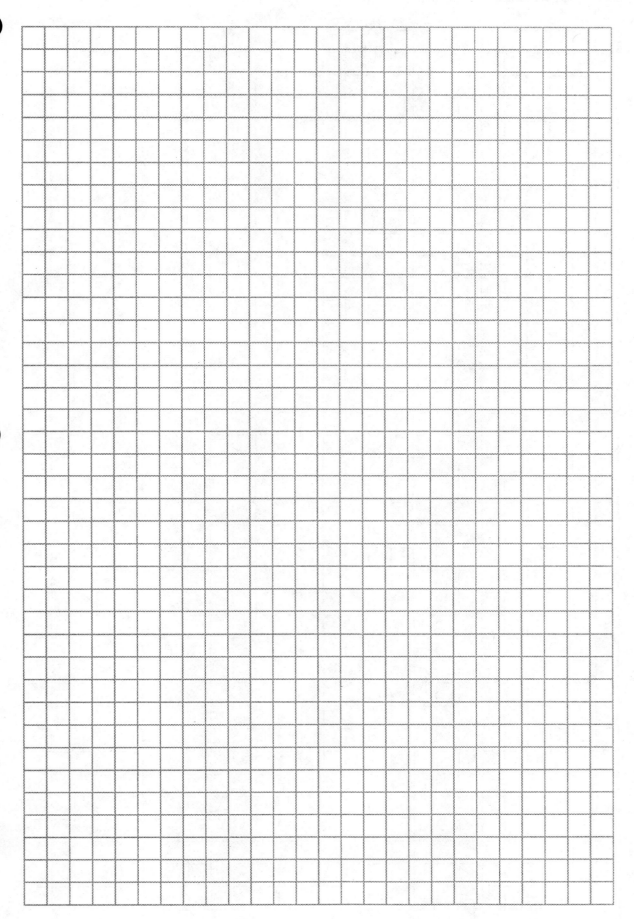

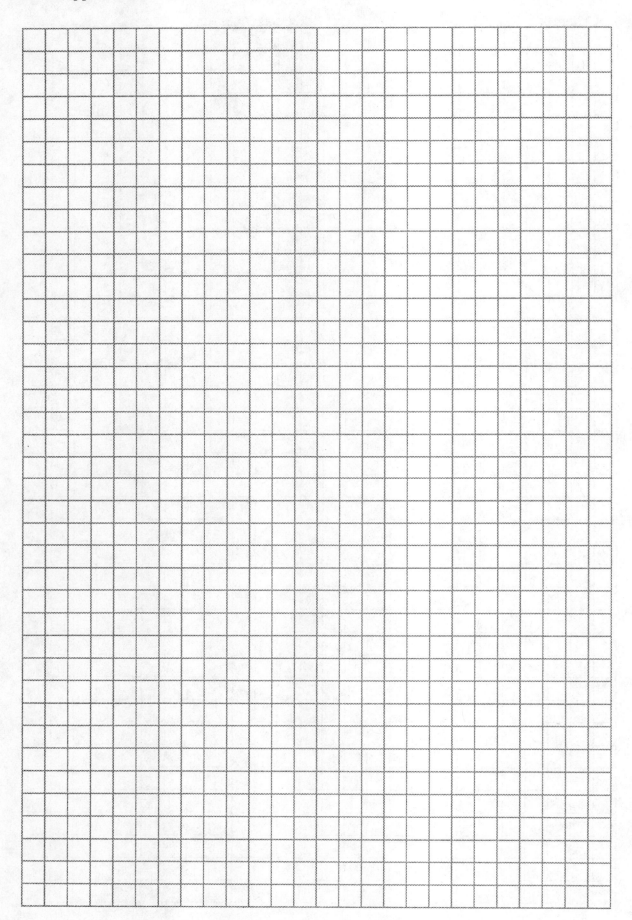

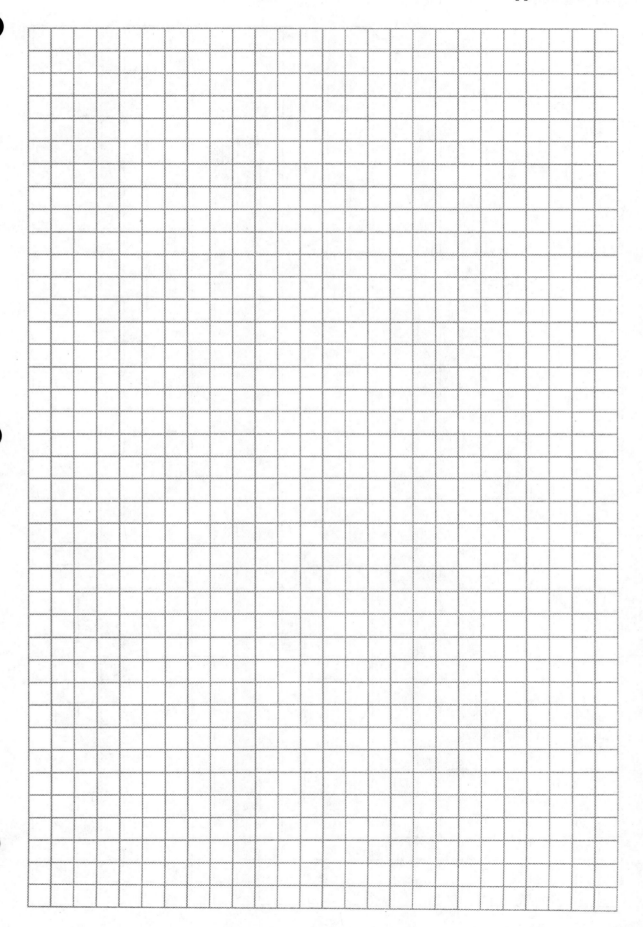

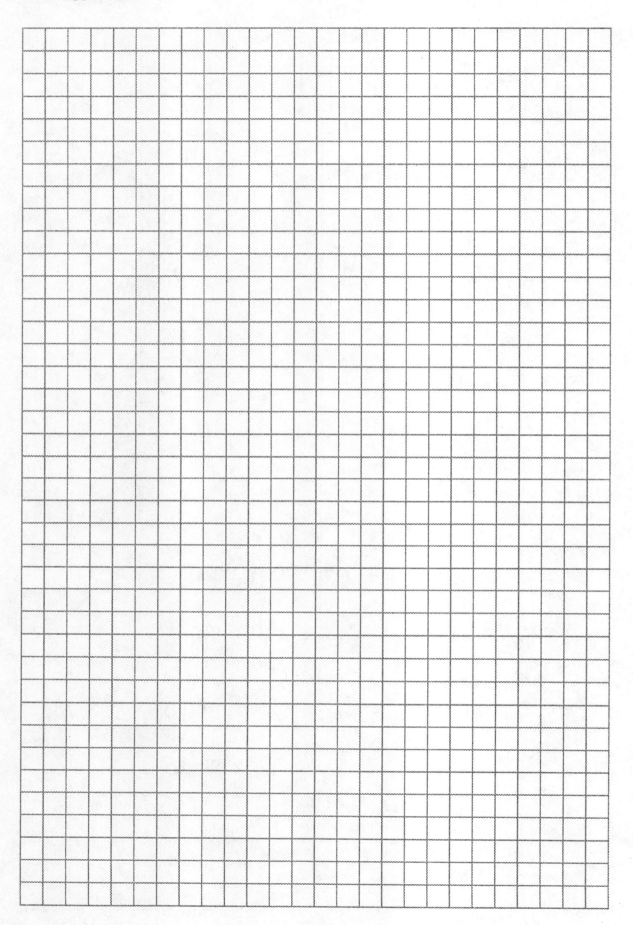

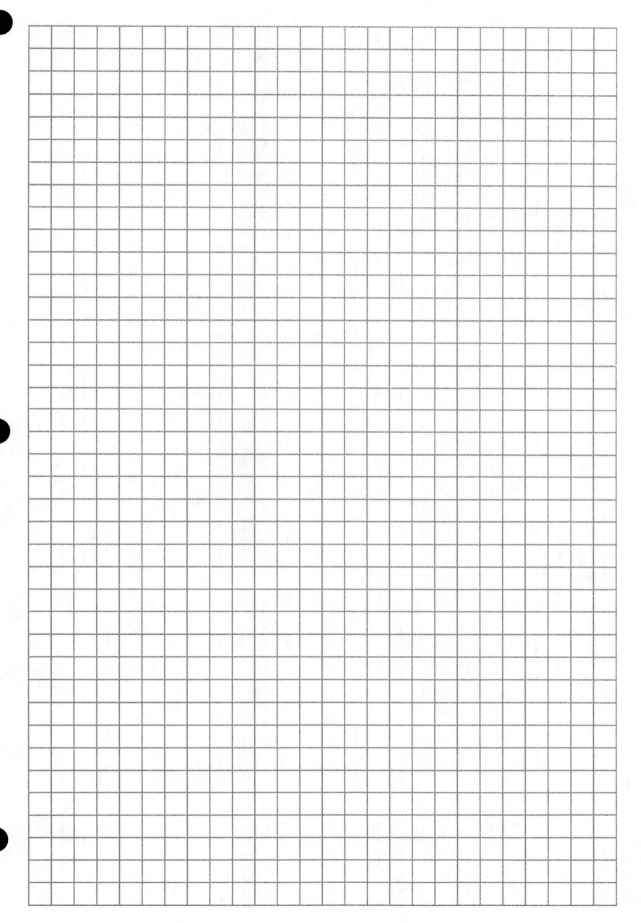

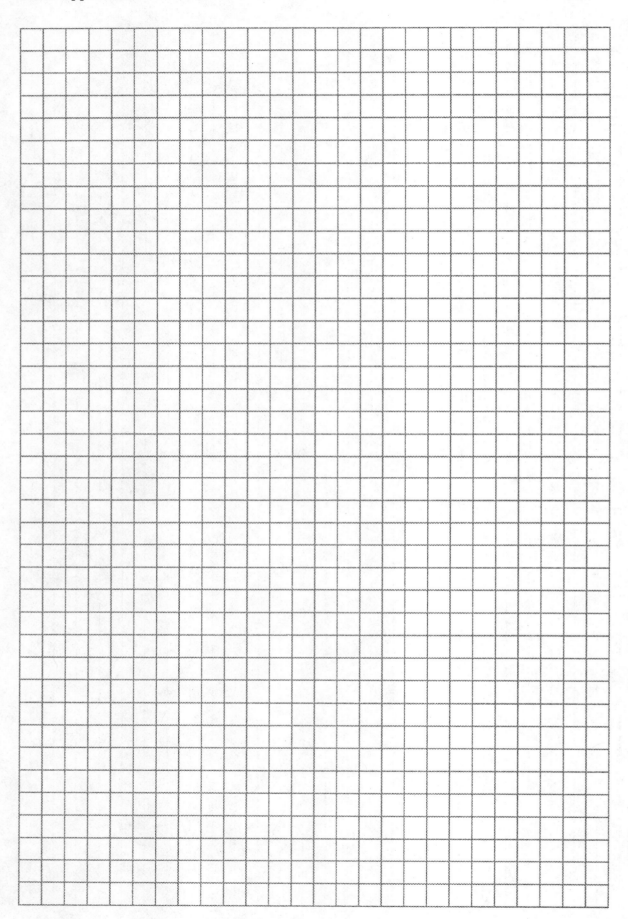

APPENDIX **B**

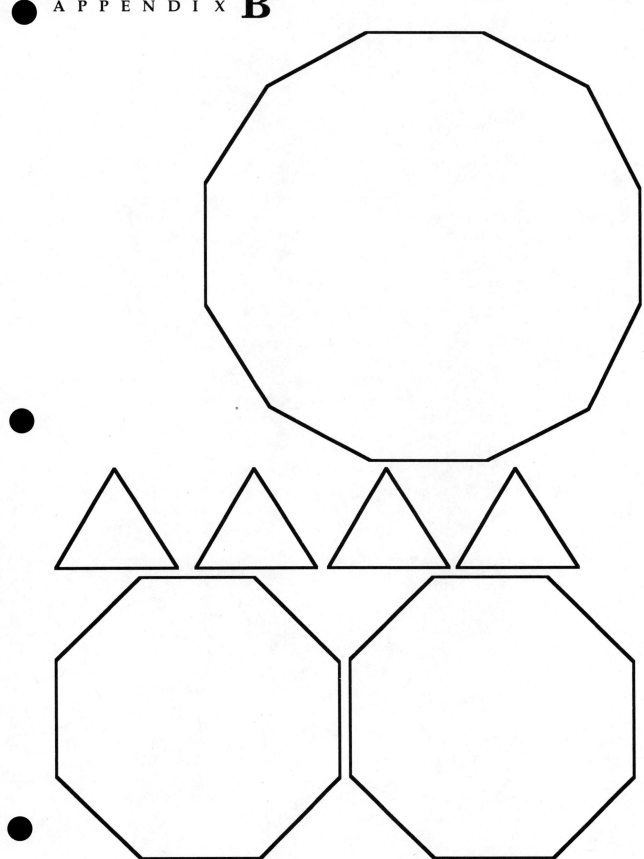

269

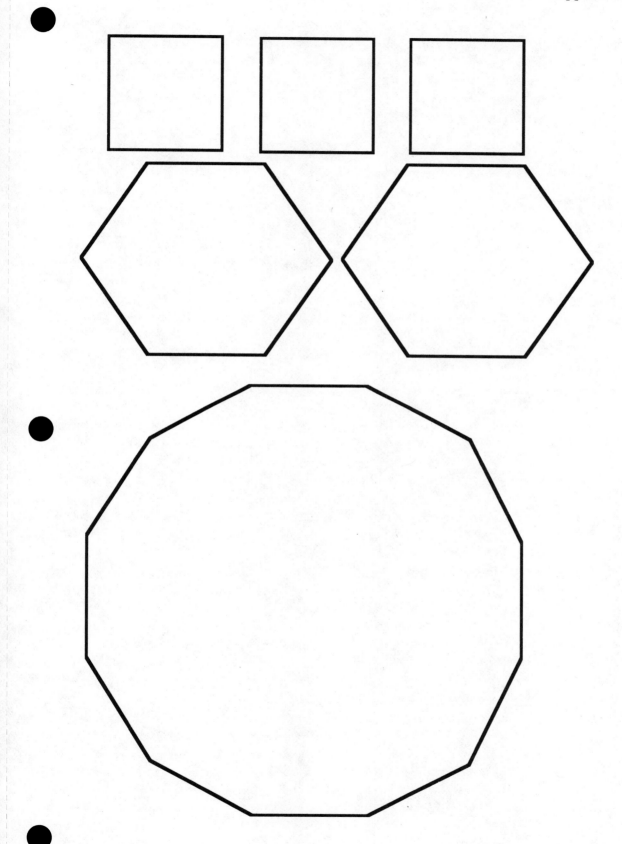

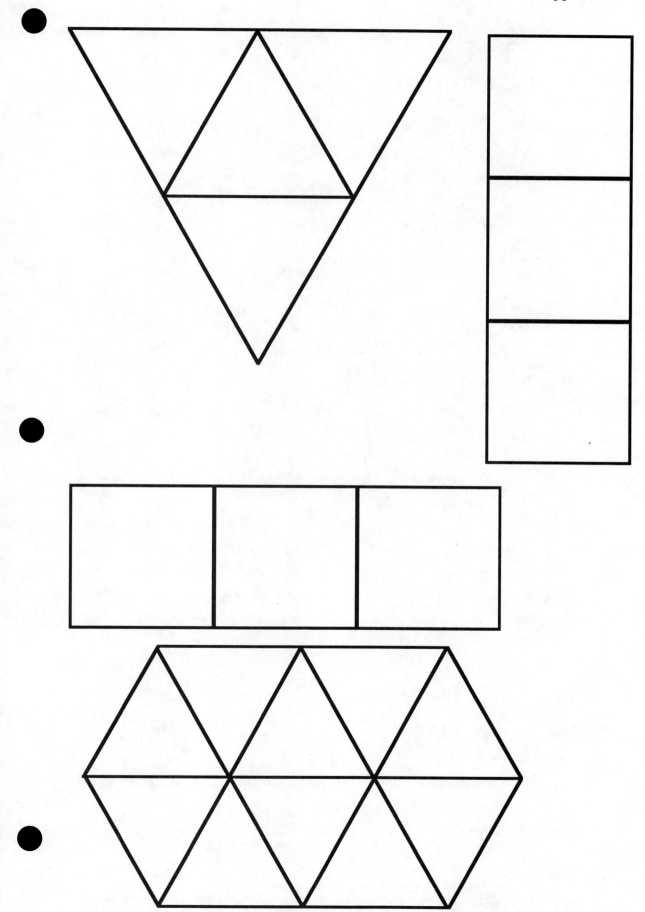

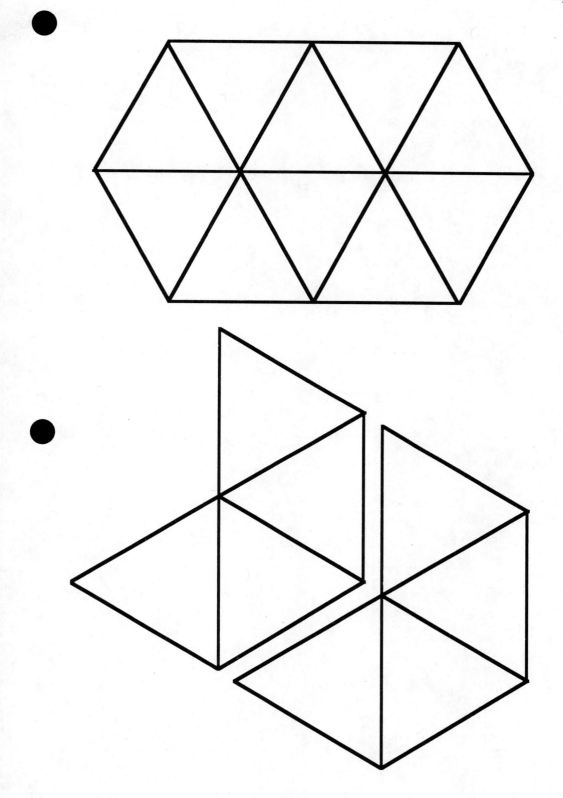

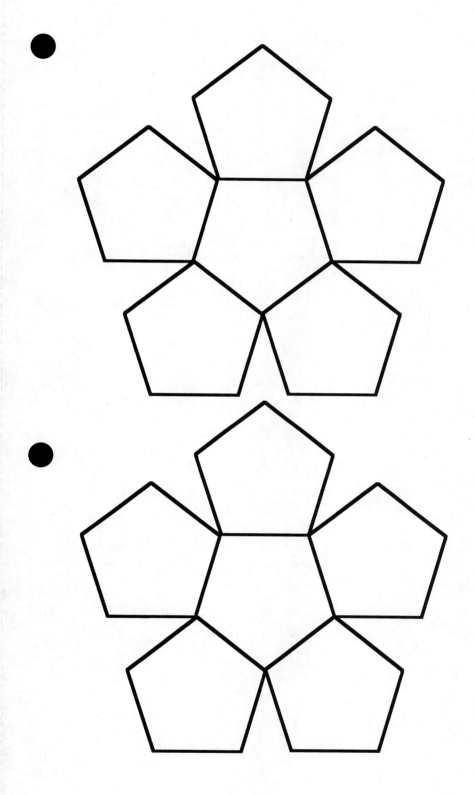

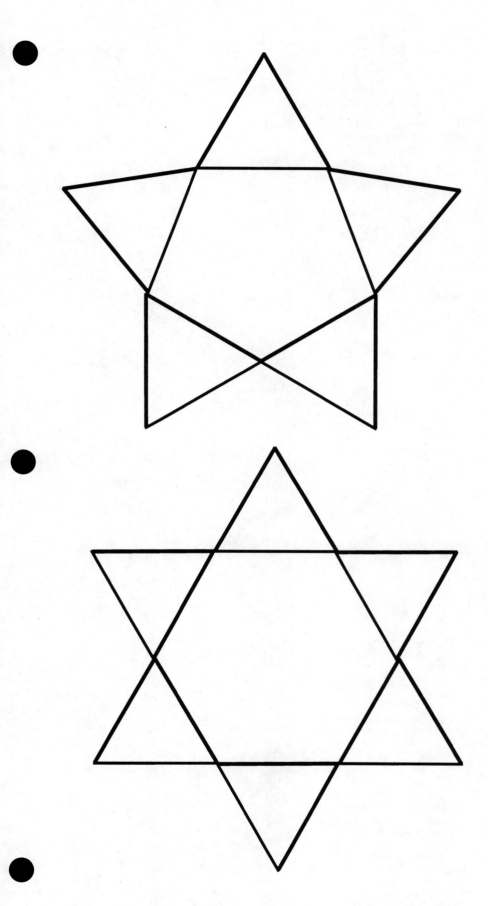

ANSWERS TO ADDITIONAL EXERCISES

Some *Using the Calculator* answers are incomplete, as they do not contain representations of the screens.

CHAPTER 1

Page 3, Supplemental Exercises

1. $\frac{3}{5}$ **2.** $\frac{4}{5}$ **3.** $\frac{5}{7}$ **4.** $\frac{24}{25}$

Page 6, Using the Calculator:

5. horizontal **6.** yes

Page 8, Reinforcing Past Lessons:

1.

Corner ball ends up in	Length	Width
Upper right	Odd	Odd
Upper left	Even	Odd
Lower right	Odd	Even
Lower left	Cannot happen	

2a. upper left, lower right **2b.** upper right

Page 11, Supplemental Exercises:

1. Answers will vary **2.** Answers will vary

Page 12, Using the Calculator:

1. 12 **2.** 8.1 **3.** 7.7 **4.** 21.1

Page 13, Supplemental Exercises:

1. $5 \times 2 + 5 \times 4$ **2.** $8 \times 4a + 8 \times 11b$ **3.** $0.5 \times 4a + 0.5 \times 12$ **4.** $12 \times 1 + 12 \times 6b$

Page 16, Supplemental Exercises:

Anthony: Guitar   Ricardo: Saxophone   Marie: Piano   David: Drums

CHAPTER 2

Page 21, Supplemental Exercises:

1a.

1	11	111	1111	11,111	111,111	1,111,111	No

1b.

$\frac{1}{2}$	$\frac{2}{3}$	$\frac{3}{4}$	$\frac{4}{5}$	$\frac{5}{6}$	$\frac{6}{7}$	$\frac{7}{8}$	No

1c.　111　　222　　333　　444　　<u>555</u>　　<u>666</u>　　<u>777</u>　　<u>Yes</u>

2.　1　　2　　4　　7　　11　　16　　<u>22</u>　　<u>29</u>　　<u>37</u>

$$+1 \quad +\underline{2} \quad +\underline{3} \quad +\underline{4} \quad +\underline{5} \quad +\underline{6} \quad +\underline{7} \quad +\underline{8}$$

Page 22, Supplemental Exercises:

　2500　<u>5000</u>　<u>7500</u>　<u>10000</u>　<u>12500</u>　15000

Page 24, Supplemental Exercises:

1.　15　　23　　31　　39　　<u>47</u>　　<u>55</u>　　<u>63</u>　　<u>71</u>

2.　t_8　is　$\underline{15} + \underline{8} \cdot \underline{7} = \underline{71}$

3a.　t_{14}　is　$\underline{32} + \underline{14} \cdot \underline{13} = \underline{214}$

3b.　t_{47}　is　$\underline{26} + \underline{21} \cdot \underline{46} = \underline{992}$

3c.　t_{88}　is　$\underline{54} + \underline{31} \cdot \underline{87} = \underline{2751}$

Page 25, Supplemental Exercises:

1. 7　**2.** 6　**3.** $\dfrac{1}{4}$

Page 27, Supplemental Exercises:

1a. $3 \cdot 2^4 = 3 \cdot 16 = 48$　**1b.** $4 \cdot \left(\dfrac{1}{2}\right)^5 = 4 \cdot \dfrac{1}{32} = \dfrac{1}{8}$　**1c.** $6 \cdot 3^3 = 6 \cdot 27 = 162$

2a. Pop.　250,000,000　<u>500,000,000</u>　1,000,000,000　<u>2,000,000,000</u>　<u>4,000,000,000</u>

2b. 2

Page 30, Supplemental Exercises:

1.	56	<u>0</u>	<u>1</u>	<u>1</u>	<u>1</u>	<u>0</u>	<u>0</u>	<u>0</u>
2.	99	<u>1</u>	<u>1</u>	<u>0</u>	<u>0</u>	<u>0</u>	<u>1</u>	<u>1</u>
3.	68	<u>1</u>	<u>0</u>	<u>0</u>	<u>0</u>	<u>1</u>	<u>0</u>	<u>0</u>
4.	12	<u>0</u>	<u>0</u>	<u>0</u>	<u>1</u>	<u>1</u>	<u>0</u>	<u>0</u>
5.	<u>16</u>	<u>8</u>	<u>4</u>	<u>2</u>	<u>1</u>	<u>27</u>		
6.	<u>64</u>	<u>32</u>	<u>16</u>	<u>8</u>	<u>4</u>	<u>2</u>	<u>1</u>	<u>126</u>
7.	<u>32</u>	<u>16</u>	<u>8</u>	<u>4</u>	<u>2</u>	<u>1</u>	<u>35</u>	
8.	<u>8</u>	<u>4</u>	<u>2</u>	<u>1</u>	<u>9</u>			

Page 32, Supplemental Exercises:

1. 15　**2.** 255　**3.** 2047

Page 36, Reinforcing Past Lessons,
1a. 34, 30, 26, 22, 18, 14, 10, 6, 2, 0   **1b.** arithmetic   **2a.** 3   6   <u>12</u>   <u>24</u>   <u>48</u>   **2b.** 2

Page 36, Using the Calculator,
1. 58   **2.** 27   **3.** 101   **4.** 424   **5.** 36,481   **6.** 9,180,900

Page 38, Supplemental Exercises,
1. 8   **2.** 7   **3.** 7   **4.** 4   **5.** 9   **6.** 9

Page 40, Reinforcing Past Lessons,
1. inductive   **2a.** 4   **2b.** 2   <u>6</u>   <u>10</u>   <u>14</u>   <u>18</u>   22

Page 43, Using the Calculator,
1.   987   <u>1597</u>   <u>2584</u>   <u>4181</u>   <u>6765</u>   <u>10946</u>   <u>17711</u>   <u>28657</u>   <u>46368</u>
2. 17711

CHAPTER 3

Page 48–9, Using the Calculator
1. y <u>−7</u>   <u>0</u>   <u>7</u>   **2.** y $-\dfrac{4}{3}$ $-\dfrac{12}{5}$ <u>−12</u>   <u>12</u> $\dfrac{12}{5}$ $\dfrac{4}{3}$ **3.** <u>3</u>   3   3

3a. A horizontal line at $y = 3$ parallel to the x axis.   **4.** y <u>9</u>   <u>4</u>   <u>0</u>   <u>4</u>   <u>9</u>

4a. no   **5.** y <u>−8</u>   <u>−1</u>   <u>0</u>   <u>1</u>   <u>8</u>   **6.** y $\dfrac{1}{128}$ $\dfrac{1}{8}$ 1   <u>4</u>   <u>8</u>

Page 52, Using the Calculator,
1a. 2   **1b.** 4   **2a.** y $-\dfrac{1}{8}$ $-\dfrac{1}{5}$ $-\dfrac{1}{3}$ <u>−1</u>   1 $\dfrac{1}{3}$ $\dfrac{1}{5}$ $\dfrac{1}{8}$

2b. There is no point on the curve for which $x = 0$.

2c. y $\dfrac{1}{8}$ $\dfrac{1}{5}$ $\dfrac{1}{3}$ 1   <u>−1</u> $-\dfrac{1}{3}$ $-\dfrac{1}{5}$ $-\dfrac{1}{8}$ **2d.** hyperbolas
2e. The curves are reflected through the x axis.

Page 55, Using the Calculator,
y <u>121.3</u>   <u>153.2</u>   <u>178.7</u>   <u>210.6</u>

Page 56, Reinforcing Past Lessons,
1a. $y = x^2 + 2$ **1b.** $y = 2x^3 + 3$ **1c.** $y = 3x + 1$ **1d.** $y = x^2 - 4$ **1e.** $y = 3x^3 + 2$

Page 58, Using the Calculator,

1a. y <u>2</u> <u>–3</u> <u>–6</u> <u>–7</u> <u>–6</u> <u>–3</u> <u>2</u>   **1b.** y <u>6</u> <u>–4</u> <u>–10</u> <u>–12</u> <u>–10</u> <u>–4</u> <u>6</u>

2a. y <u>–11</u> <u>5.56</u> <u>27.64</u> <u>35</u> <u>27.64</u> <u>5.56</u> <u>–11</u>

Page 59–60, Using the Calculator,

1a. y <u>18</u> <u>3</u> <u>2</u> <u>3</u> <u>18</u>   **1b.** y <u>–29</u> <u>2</u> <u>3</u> <u>4</u> <u>35</u>

Page 61–2, Reinforcing Past Lessons,

1a. geometric   **1b.** 5   **1c.** 87

1d. $3 + 7 \times 12$; The first term plus the common difference times the number of term asked for minus one, equals 87.

1e. 11001 <u>25</u>   110 <u>6</u>

   1010 <u>10</u>   11111 <u>31</u>

1f. 27 <u>11011</u>   33 <u>100001</u>

   49 <u>110001</u>   104 <u>1101000</u>

1g. 25, 36, 49, 64, 81   **1h.** 3   **1i.** 9642 and 89642   **1j.** 64, 125, 216, 343, 512

1k. 5, 8, 13, 21, 34   **2a.** y <u>2.6</u> <u>4.55</u> <u>10.4</u> <u>14.3</u> <u>18.2</u>

2b. They are close, but slightly smaller.

Page 64, Supplemental Exercises:

1a.

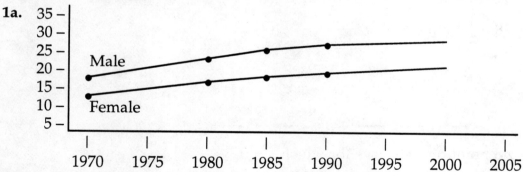

1b. males 21.35, females 15.4, interpolate   **1c.** males 28.4, females 21.5, extrapolate (Remember that since you are extrapolating, the answer might not be exact.)

Page 65, Supplemental Exercises:

1.

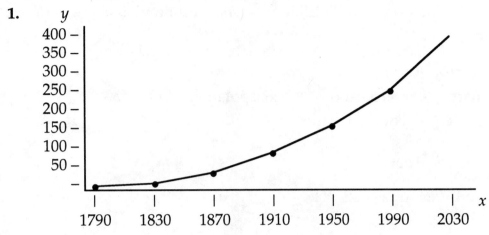

1b. around 390, extrapolate

Page 65–6, Using the Calculator:

1.

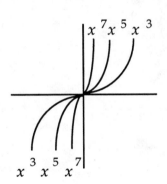

2.

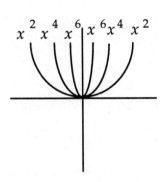

4. There is no point on the curve for which $x = 0$.   **5.** It is impossible to find a value for y if $x = 0$ because $\dfrac{5}{0}$ is not equal to any number.

6. y $\underline{-.6}$ $\underline{-1.25}$ $\underline{-5}$ $\underline{\text{no number}}$ $\underline{5}$ $\underline{1.25}$ $\underline{.6}$ **8.** y $\underline{-4}$ $\underline{-3}$ $\underline{4}$ $\underline{23}$ $\underline{60}$ $\underline{121}$

Page 67, Supplemental Exercises:

1a. y $\underline{5}$ $\underline{8}$ $\underline{17}$ $\underline{32}$ $\underline{53}$ $\underline{80}$ **1b.** y $\underline{-1}$ $\underline{0}$ $\underline{7}$ $\underline{26}$ $\underline{63}$ $\underline{124}$ **2a.** $y = 2x^3 + 4$ **2b.** $y = 12x + 1$

CHAPTER 4

Page 70 Using the Calculator

1. 10^9 (Answers will vary depending on calculator used.)   **2.** E24 (Answers will vary   **5.** Answers will vary   **6.** The curve geets closer to the x-axis as it moves left, but never reaches it.

Page 74, Supplemental Exercises:

1. 2,516,000,000   **2.** 6,261,000,000   **3.** 2.9×10^4

Page 76, Reinforcing Past Lessons,

1. 3 nonillion   **2.** 17 septillion   **3.** 132 trillion   **4.** 435 decillion   **5.** 50 sextillion
6. 194 quadrillion

Page 81–2, Reinforcing Past Lessons,

1. decimal   **2.** binary   **3.** interpolation   **4.** extrapolation
5. common difference   **6.** common ratio

Page 87 Using the Calculator

3. y <u>3.2</u>   <u>3.3</u>   <u>4.1</u>   <u>6.1</u>   <u>9.8</u>   <u>24.5</u>   <u>70.9</u>   <u>99.7</u>

CHAPTER 5

Page 92, Supplemental Exercises

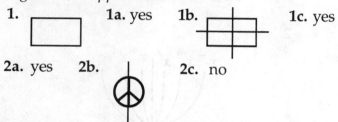

1.   **1a.** yes   **1b.**   **1c.** yes

2a. yes   **2b.**   **2c.** no

3a. Gemini, Libra, Sagittarius, Aquarius, Pisces, Aries, and Taurus
3b. Gemini, Cancer, Pisces

Page 94, Supplemental Exercises
1. and 2. Answers will vary. **3a.** B, E, F, G, I, K   **3b.** A, C, D, J, K

Page 94, Reinforcing Past Lessons,
1. 5.201×10^9 **2.** 5.38×10^8 **3.** 6.13×10^8, $1,036 \times 10^9$

Page 98, Reinforcing Past Lessons,
1. hexagon   **2.** pentagon   **3.** square   **4.** decagon   **5.** heptagon   **6.** dodecagon

Page 100 Using the Calculator

y $3\frac{5}{8}$ $3\frac{2}{5}$ <u>1</u>   <u>–1</u>   <u>–3.5</u>   <u>no number</u>   <u>11.5</u>   <u>9</u>   <u>7</u> $4\frac{3}{5}$ $4\frac{3}{8}$

Page 101, Reinforcing Past Lessons,
1. 8   **2.** 3×8^7; 6,291,456   **3a.** <u>34</u>   <u>55</u>   <u>89</u>; Fibonacci   **3b.** <u>36</u>   <u>49</u>   <u>64</u>; sequence of
squares   **3c.** <u>125</u>   <u>216</u>   <u>343</u>; sequence of cubes   **4.** 5   **5.** 1101011   **6.** 2401
7. three octillion   **8.** 10^6; 1000000

Page 103–4, Using the Calculator

1. *y* <u>−9</u> <u>4</u> <u>5</u> <u>0</u> <u>−5</u> <u>−4</u> <u>9</u>

Page 104–5, Reinforcing Past Lessons

1. $\dfrac{360}{n}$ **2.** *y* <u>45</u> <u>36</u> <u>30</u> <u>20</u> <u>18</u> **3.** 15 feet **4.** regular polygon **5.** tetrahedron

6. octahedron **7.** icosahedron **8.** 32 **9.** 100,000,000

Page 111–2, Using the Calculator

1a. Yes **1b.** Yes; answers will vary **2a.** Yes **2b.** No **3a.** Yes **3b.** Yes; 4-fold or 16-fold **4a.** Yes **4b.** No

CHAPTER 6

Page 113, Reinforcing Past Lessons

1a. *y* <u>−21</u> <u>−18</u> <u>−15</u> <u>−12</u> <u>−9</u> <u>−6</u> <u>−3</u> <u>0</u> **2.** *y* <u>−2</u> <u>3</u> <u>6</u> <u>7</u> <u>6</u> <u>3</u> <u>−2</u> **2a.** parabola

Page 116 Supplemental Exercises

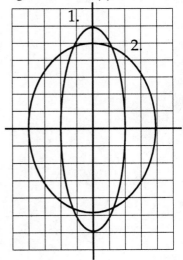

Page 123 Supplemental Exercises

1. 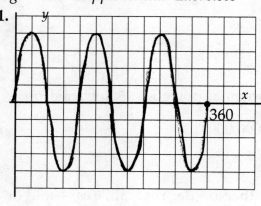   **2.**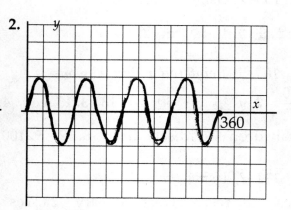

Page 124, Reinforcing Past Lessons
1a. three   **2a.** six

Page 127, Supplemental Exercises
1.   1.8   3.0   4.2   <u>5.4</u>   <u>6.6</u>   <u>7.8</u>   **1a.** arithmetic   **1b.** Archimedean

Page 128 Using the Calculator
1a. Archimedean

Page 130, Using the Calculator
1a. a special type of cycloid

Page 132, Reinforcing Past Lessons
1.

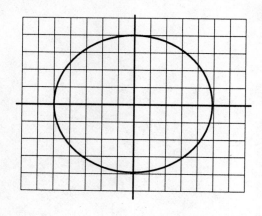

1a. ellipse

CHAPTER 7

Page   134,   Supplemental   Exercises

First Question	Second Question	Third Question	Outcomes

```
                                    T        TTT
                          T
                                    F        TTF
                                    T        TFT
          True
                          F
                                    F        TFF

                                    T        FTT
                          T
                                    F        FTF
                                    T        FFT
          False
                          F
                                    F        FFF
```

1a. $2^3 = 8$ **2.** $10 \times 10 \times 26 \times 10 \times 10 \times 10 \times 10 = 26{,}000{,}000$

3. $26 \times 26 \times 10 \times 10 \times 10 \times 10 = 6{,}760{,}000$ **4.** $7 \times 7 \times 7 \times 10 \times 10 \times 10 \times 10 = 3{,}430{,}000$

Page   136,   Supplemental   Exercises

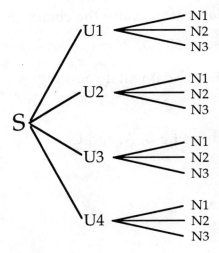

1a. $4 \times 3 = 12$

Page 138, Reinforcing Past Lessons

1.

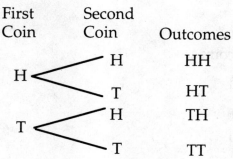

1a. $2 \times 2 = 4$

Page 140, Supplemental Exercises

1a. no, because $10! = 9! \times 10$, not $9! + 1!$ **1b.** yes, because $3! = 6$, and $6! = 5! \times 6$
1c. yes, because $7! = 6! \times 7$ and, $6! = 5! \times 6$ **2.** $6! = 720$ **3.** $4! = 24$ **4.** $10 \times 9 = 90$
5. $8! = 40,320$ **6.** $3! \times 3! = 36$

Page 142, Supplemental Exercises

1. $\dfrac{5!}{2! \times 2!} = \dfrac{120}{4} = 30$ **2.** $\dfrac{7!}{3!} = \dfrac{5040}{6} = 840$

Page 145, Supplemental Exercises

1a. $8 \times 7 \times 6 \times 5 = 1680$ **1b.** permutation   **1c.** order matters since the chairs are
different   **2.** $\dfrac{11 \times 10 \times 9 \times 8 \times 7 \times 6 \times 5 \times 4}{8!} = \dfrac{11 \times 10 \times 9}{3!} = 165$ **3.** $2^2 = 4$
3a. $55, 56, 65, 66$ **4.** $5^5 = 3125$ **5.** $5 \times 3 \times 2 = 30$ **5a.** fundamental counting
principle   **6.** $\dfrac{22 \times 21 \times 20}{3!} = 1540$
6a. no; choosing books A, B, and C is the same as choosing B, A, and C
6b. combination   **7.** $\underline{2}\ \underline{4}\ \underline{8}\ \underline{16}$ **7a.** independent
7b. fundamental counting principle

CHAPTER 8

Page 150, Supplemental Exercises
1a. $\dfrac{6}{11}$ **1b.** $\dfrac{3}{11}$ **1c.** $\dfrac{2}{11}$ **2.** $\dfrac{1}{5}$; 20%

Page 151, Supplemental Exercises
1. $\dfrac{7}{36}$; 19.4%   **2.** $\dfrac{1}{4}$; 25%

Page 154, Reinforcing Past Lessons:

1.

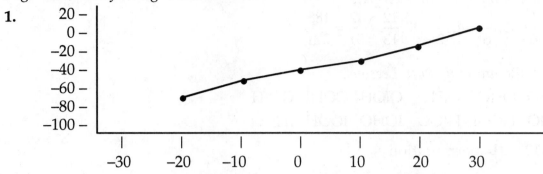

1a. −80   **1b.** −31   **1c.** −3   **1d.** b. and c.   **1e.** a.   **2.** $5 \times 6 \times 4 = 120$

Page 157, Supplemental Exercises

1a. $\dfrac{1}{6} \times \dfrac{1}{6} = \dfrac{1}{36}$ **1b.** $2.\overline{7}\%$ **2a.** $\dfrac{1}{13} \times \dfrac{1}{17} = \dfrac{1}{221}$ **2b.** .452%   **3a.** $\dfrac{1}{2} \times \dfrac{25}{51} \times \dfrac{12}{25} = \dfrac{2}{17}$

3b. 11.8%   **3c.** dependent   **4a.** $\dfrac{3}{13} \times \dfrac{11}{51} \times \dfrac{1}{5} = \dfrac{11}{1105}$ **4b.** .995%   **4c.** dependent

5a. $\dfrac{1}{4} \times \dfrac{4}{17} \times \dfrac{11}{50} = \dfrac{11}{850}$ **5b.** 1.29%   **6a.** $\dfrac{3}{13} \times \dfrac{3}{13} \times \dfrac{3}{13} = \dfrac{9}{2197}$ **6b.** .41%

6c. independent   **7a.** $\dfrac{5}{6}$ **7b.** $83.\overline{3}\%$

Page 161, Supplemental Exercises

1. $.420 \times .420 \times .420 = .074$ **2.** $.376 \times .376 \times .376 \times .376 = .020$

3a. $389 \times .389 \times .389 \times .389 = .023$ **3b.** $389 \times .389 \times .389 \times .389 \times .389 = .009$ **4.** $\dfrac{5}{6}$

Page 166, Supplemental Exercises

1.

No. of boys	0	1	2	3	4	5	6	7
No. of ways	1	7	21	35	35	21	7	1
Fractional probability	$\dfrac{1}{128}$	$\dfrac{7}{128}$	$\dfrac{21}{128}$	$\dfrac{35}{128}$	$\dfrac{35}{128}$	$\dfrac{21}{128}$	$\dfrac{7}{128}$	$\dfrac{1}{128}$
% probability	.8%	5.5%	16.4%	27.3%	27.3%	16.4%	5.5%	.8%

1a. 77.3%   **1b.** .8%   **1c.** 50 percent

2.

No. of tails	0	1	2	3	4	5	6	7	8	9
No. of ways	1	9	36	84	126	126	84	36	9	1
Fractional probability	$\dfrac{1}{512}$	$\dfrac{9}{512}$	$\dfrac{18}{256}$	$\dfrac{21}{128}$	$\dfrac{63}{256}$	$\dfrac{63}{256}$	$\dfrac{21}{128}$	$\dfrac{18}{256}$	$\dfrac{9}{512}$	$\dfrac{1}{512}$
% probability	.2%	1.8%	7%	16.4%	24.6%	24.6%	16.4%	7%	1.8%	.2%

2a. $\dfrac{1}{2}$ **2b.** .2%   **2c.** $\dfrac{65}{256}$

Page   169, Supplemental Exercises

1. 1   **2.** $\dfrac{1}{6}$ **3.** $\dfrac{5}{6}$ **4.** $\dfrac{1}{2}$ **5.** $\dfrac{12}{13} \times \dfrac{47}{51} = \dfrac{188}{221}$

Page 169, Reinforcing Past Lessons:

1.   OHIO   OHOI   OIHO   OIOH   OOHI   OOIH
   HOIO   HOOI   HIOO   IOHO   IOOH   IHOO

1a. $\dfrac{4!}{2!} = 12$ **1b.** permutation

Page 170, Using the Calculator:

1. y <u>16</u> <u>14</u> <u>12</u> <u>10</u> 8 6   **2a.** ellipse

CHAPTER 9

Page 177, Supplemental Exercises

1a. Frequency: 1, 2, 3, 3, 1

1b.

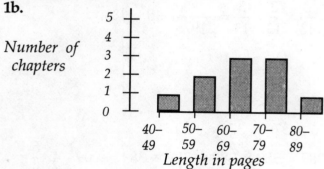

Page 178, Using the Calculator

1a. No   **1b.** ellipse   **1c.** $\dfrac{x^2}{4} + \dfrac{y^2}{16} = 1$

Page 181–2, Reinforcing Past Lessons:

1.

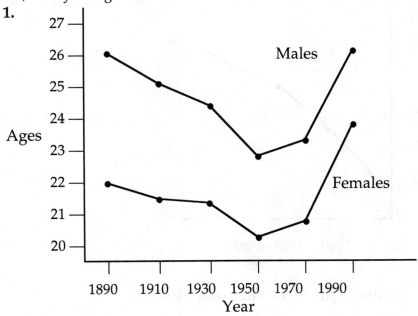

1a. Answers will vary.   **2a.** $2 \times 2 \times 2 \times 2 \times 2 \times 2 = 2^6 = 64$ **2b.** fundamental counting principle   **3.** $_4P_4 = 4 \times 3 \times 2 \times 1 = 4! = 24$ **4.** $_{11}P_3 = 11 \times 10 \times 9 = 990$

5. y

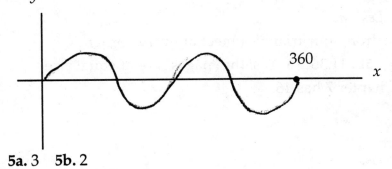

5a. 3   **5b.** 2

Page 185, Supplemental Exercises

1a. 65   **1b.** 62   **1c.** 62

Page 185–6, Reinforcing Past Lessons

1.

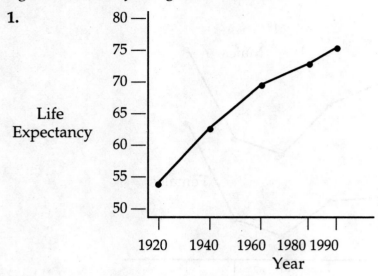

1a. 66.3   **1b.** interpolation   **1c.** 77.1   **1d.** extrapolation

Page 192, Using the Calculator
1a. yes   **1c.** yes; line symmetry

Page 197–8, Reinforcing Past Lessons
1. 13   **2.** 67,108,864   **3.** octahedron; four triangles meet at every corner
4. biomial probability   **5a.** 36   **5b.** 11.36   **6.** Yes; the "pages" are of equal size
and Chapter 5 has 82, while Chapter 7 has 46.

CHAPTER 10

Page 201, Supplemental Exercises
1. B, E   **2.** C   **3.** D   **4.** none

Page 202, Using the Calculator
1a. no   **1b.** ellipse   **1d.** No; the spacing for numbers on the calculator is
different on the x– and y–axes.   **1e.** circle   **1f.** Now the spacing is equal for each
number interval.

Page 209–10, Using the Calculator
1a. y .49 –5.12 –7 –5.12 .49 9.85   **1c.** y .1 –5 –7 –5 1 11   **1d.** parabola
1e. yes; one (vertical)   **1f.** no   **1g.** two

Page 201, Supplemental Exercises
1a. yes   **1b.** 12   **1c.** 11   **1d.** 5   **1e.** White River all the way. (Answers will vary.)

Page 212, Using the Calculator

1b. 7   **1c.** about four   **1d.** $\dfrac{360}{1.5} = 240$ **1e.** sine curve